THE SCIENCE OF MIND

First MIT Press edition, 1989
© 1989 by Editions Hologramme

French edition published by Editions Hologramme, France, under the title
Les Énigmes du cerveau.

World rights in English are held by The MIT Press.

All rights reserved. No part of this book may be produced in any form by any electronic or mechanical means (including photocopying, recording, or information storage and retrieval) without permission in writing from the publisher.

This book was set by The MIT Press and printed and bound by Dai Nippon in Japan.

Library of Congress Cataloging-in-Publication Data

Klivington, Kenneth A.
The science of mind/Kenneth A. Klivington; scientific advisors Floyd Bloom... (et al.).

p. cm.
ISBN 0-262-11141-1
1. Brain.   2. Neurophysiology.
3. Neuropsychology.   I. Title.
[DNLM: 1. Brain–physiology.
2. Mental Processes–physiology.
WL 300 K657]
QP376.K49513 1989
152–dc 19
DNLM/DLC
for Library of Congress   89-2733
                                        CIP

THE SCIENCE OF MIND

Kenneth A. Klivington

Illustrations selected by Elizabeth Antébi

Scientific advisors

Floyd Bloom
Member of the Research Institute of Scripps Clinic

Valentino Braitenberg
Director of the Max Planck Institute for Biological Cybernetics

Jacques Glowinski
Professor at the Collège de France

Roger Guillemin
Distinguished Scientist, Whittier Institute for Diabetes and Endocrinology, Nobel Prize in Physiology or Medicine in 1977

David Hubel
Professor at Harvard Medical School, Nobel Prize in Physiology or Medicine in 1981

Masao Ito
Professor at the University of Tokyo

Leslie Iversen
Director of the Neuroscience Research Center at Merck, Sharp, and Dohme Research Laboratories

Wolf Singer
Director of the Max Planck Institute for Brain Research

The MIT Press
Cambridge, Massachusetts
London, England

Contents

	Introduction	6
PART 1	**THE DIMENSIONS OF MIND**	9
	Chapter 1. Consciousness: A Matter of Thought	11
	Article 1. Brain Landscapes of Consciousness, Language, and Dementia David H. Ingvar	16
	Chapter 2. Brain Power: The Search for Intelligence	21
	Article 2. The Brain and Evolution Yves Coppens	26
	Article 3. Lemurs, Ancestors of Our Ancestors Véronique Barre and Jean-Jacques Petter	28
	Chapter 3. The Minds of Animals: Behaviorism and Ethology	33
	Article 4. Cricket Love Behavior and the Cricket Brain Franz Huber	39
	Chapter 4. Gender, Culture, and Genes Make a Difference	45
	Article 5. The Brain and the Medical Traditions of the Eastern World Jean-Claude de Tymowski	51
	Article 6. Hemispheric Dominance in Japan and the West Tadanobu Tsunoda	54
	Article 7. Sex Differences in Human Brain Organization Doreen Kimura	60
	Chapter 5. It Takes Timing: The Rhythms of Life	63
	Article 8. The Biological Clock Alberto Oliverio	71
	Article 9. Sleep and Dreaming Allan Hobson	74
	Article 10. Alterations in Brain Function during Weightlessness Laurence R. Young	78
	Chapter 6. Emotions: A Gut Feeling	81
	Chapter 7. Strange Thoughts: The Disordered Mind	85
	Article 11. Understanding and Treating Anxiety W. Haefely, J. G. Richards, and H. Möhler	94
	Article 12. Senile Dementia and the Acetylcholine Connection J. M. Palacios	98
	Article 13. Toxins as Keys to Brain Communication James W. Patrick	102
PART 2	**THE DIMENSIONS OF THE BRAIN**	105
	Chapter 8. Paths through the Brain: Maps for Explorers	107
	Chapter 9. Is It Real? What Our Senses Tell Us	117
	Article 14. The Neural Basis of Olfaction Eric Barrington Keverne	125
	Article 15. There Is More to the Smell of Pigs Than Meets the Nose W. D. Booth	128
	Article 16. Hearing: A Collaboration of Brain and Ears Nelson Yuan-shen Kiang	130
	Chapter 10. Who's in Charge? The Control of Movement	135
	Article 17. Brain and Motor Control Emilio Bizzi	142
	Article 18. The Roles of the Cerebellum and Basal Ganglia in Motor Control Masao Ito	144

	Chapter 11. Brain Development: A Matter of Life and Death	147
	Article 19. Cell Death in the Developing Nervous System Max Cowan	152
	Chapter 12. The Brain's Chemistry Set	155
	Article 20. The Chemical Language of Interneuronal Communication Floyd Bloom	160
	Article 21. Hormones of the Brain Roger Guillemin	162
	Article 22. "Morphines" of the Brain Lars Terenius	164
	Article 23. Brain Implants Anders Björklund	167
	Chapter 13. The Stuff Memories Are Made Of	171
	Article 24. The Brain: A Self-Organizing Learning System Wolf Singer	174
PART 3	MIND-BRAIN CONNECTIONS	181
	Chapter 14. Thinking Healthy Thoughts: Are You What You Think?	183
	Article 25. Paradoxes of Depression Pierre Pichot	190
	Article 26. Is the Brain Involved in the Elaboration of Immune Responses? Kathleen Bizière and Gérard Renoux	192
	Chapter 15. New Knowledge from Damaged Brains	195
	Article 27. The Interpretive Brain Michael S. Gazzaniga	203
	Article 28. Electrical Signs of Lateral Dominance in the Human Brain Richard Jung	205
	Chapter 16. Speaking Your Mind: What Is Language?	209
	Article 29. Developmental Dyslexia Albert M. Galaburda	212
	Article 30. A Drug for Dyslexia C. K. Conners	216
	Chapter 17. Focusing In: Attention and the Brain	219
	Chapter 18. Thought as Computation: The Future of Brain Research?	223
	Article 31. Computers and Brains Tomaso Poggio	226
	Article 32. Brains and Computers Valentino Braitenberg	229
	Conclusion	232
	Notes	235
	Suggested Readings	237
	Index	238

Introduction

Nothing is more complicated than the human brain. And how the human brain works is the greatest scientific puzzle that we know. There are, of course, many other challenging scientific questions that remain to be answered. These include such admittedly difficult problems as how cancer occurs and how the universe took shape. But for these questions, scientists have at least an idea of the fundamental nature of the problem, know in general where to look for answers, and know what sort of answers to expect. How the brain works is a problem of an altogether different sort. This becomes clear when we try to imagine how someone might possibly answer the deceptively straightforward question, How does the brain work?

The first difficulty occurs when we try to decide precisely what it is we would like to explain about the brain. Is it some rather abstract property generally associated with the brain like consciousness or intelligence? Or is the goal to explain how more specific functions like learning and memory storage are carried out? It might seem an even more straightforward matter to ask how our brains interpret information from our eyes, ears, or other sense organs. Does the brain work like a computer, receiving information, processing it, storing it, and producing output? How does a brain cell work? How are brain cells connected together? What happens to our brains when we get old? Do brain cells grow back again after they are damaged?

This is only a small sample of the more specific questions we might list when we seriously ask how the brain works or how it fails to work as a consequence of damage, disease, or age. But even when we have settled on the questions we would like to ask, our problems are only beginning. We must next decide what kind of answer to look for. Do we want to know about chemical reactions in brain cells, or are we in search of something like a computer program that explains how we can recognize a face or a voice? Will it be sufficient simply to identify the parts of the brain that carry out a specific task, or do we want to identify every brain cell involved and specify every detail of its activity? Do we need a psychologist to answer our questions, or a molecular biologist or both?

The purpose of this book is to help resolve dilemmas such as these: to clarify the kinds of sensible questions that can be asked about the functions of the brain and to scan the vast field of modern brain research to see what answers are available now or are likely to come in the future. Explanations of how the brain works tend to be dressed in the fashions of current technology. Not too many years ago people talked about the brain as a kind of telephone exchange. Messages come from the eyes, ears, nose, and other sense organs and get shunted through the central exchange of the brain. From there, once the right connections are made, outgoing messages get delivered to the correct muscles, producing an appropriate response. This was a useful but very limited analogy. What happened in the "central exchange" of the brain remained a mystery.

More recently, the computer has become the technology of choice to explain the brain. Information processing psychology relates what people do to how machines process information. Brains receive information from the environment, process it, store memories, produce outputs of movement and language, and do other things that make them roughly comparable to computers, but only roughly. It is increasingly clear that brains are not really like computers, at least not the common type of computer known today. Computers under development seem to function somewhat more like biological brains in that they can do more than one thing at a time. They have the capability to do something known as parallel processing; they operate rather like a collection of separate computers interconnected with one another, all working on separate but related parts of a problem. But it seems that

Introduction

computer scientists are learning more about how to build these kinds of novel machines from brain research than neuroscientists are learning about brains by studying computers.

This is not a book about the brain for readers interested in doing research, so it does not contain all the details to be found in a textbook. There are no technical descriptions of the intricate details of the electrical properties of nerve cells. These are fascinating in themselves, but they are in the realm of academic treatises. This is a book for readers who seriously want to learn what we know about the brain, but do not care to absorb all the complex details of neuroanatomy, neurophysiology, neurochemistry, and other subjects that appear in scholarly works. So the material that appears here, while adhering to what is scientifically correct, does not follow the traditional pattern of books about the brain. It is concerned instead with questions that are likely to be important to readers who are curious about this most vital of scientific frontiers. As the reader will discover, it is a book that really has more to do with questions than with answers.

It is not realistic to try to explore all the possible kinds of questions and answers about the brain in a book of this size. Instead, I shall be selective, choosing first to explore some of the questions about behavior and thought that arouse most everyone's curiosity. These include matters of consciousness, emotions, and mental illness. In part 1, I shall examine properties of the human mind that seem to have some relation to the biology of the brain. While some satisfying answers will emerge, the effort underscores the difficulty of the problems philosophers have wrestled with for centuries in trying to relate the mind to the brain. Like all muses, those of the brain often respond to questions with answers that are more baffling than illuminating.

Part 2 turns toward the more biological side of the mind-brain problem by exploring some of the answers the biologists have come up with in asking how the brain's basic machinery operates. These answers involve some details of rather basic biology and chemistry, but they can be understood without the need for any special training in these subjects. The intent is to provide a general sense of the kinds of answers that are now available as a result of modern developments in molecular and cellular biology and many other branches of science involved in the study of the brain. It is by no means a complete account, but it does provide a representative sample of the kinds of things we currently know about what gets into the brain, what happens to it there, and what comes out.

In part 3, I delve into rapidly developing new research that is beginning to reveal a few of the underlying connections between mind and brain. This is occurring in such new developments as the emergence of psychoneuroimmunology, which is concerned with biological connections that might account for how states of mind alter the effectiveness of the body's immune system. Other research on language, one of the basic tools of the mind, is showing how language processing may be a built-in capability of the human brain. Finally, research that crosses into the realm of artificial intelligence explores ways in which the mind's operations may be viewed as computations performed by the brain.

The ancient Greeks knew full well the importance of asking the right questions from their dealings with oracles and muses, whose answers could be so bewildering. Today we must be equally careful with the muses of the brain, particularly since there seem to be so many of them. They still keep from us many of the brain's deepest secrets. But at least we seem to be learning how to ask the right questions.

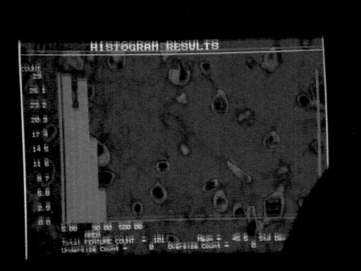

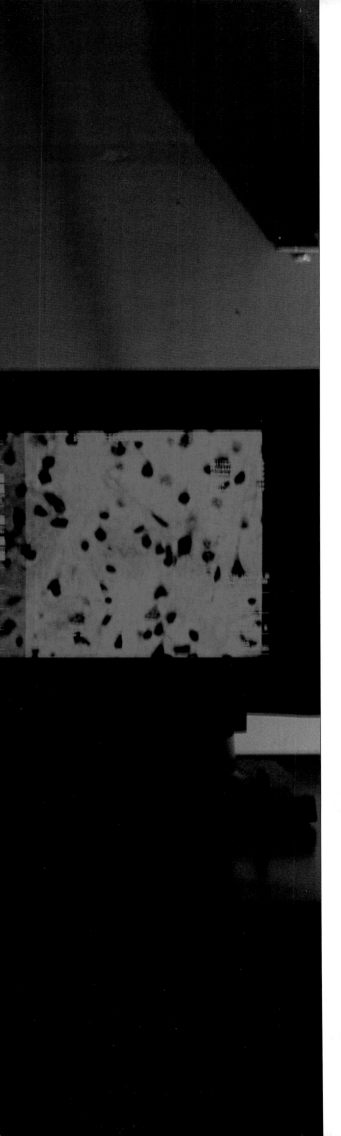

I
THE DIMENSIONS OF MIND

If we do not know what questions to ask about the brain and have no idea of what kinds of answers we would like to have, an exploration of brain research will be fruitless and bewildering. Miles of library shelves are filled with books and journals packed with detailed technical information about the functions of the brain, from the chemicals it contains to the treatment of the mentally ill. Yet for all this information, our present position is one of almost total ignorance.

My purpose is not to condense the great masses of technical detail, which, for all their bulk, can leave a reader frustratingly uninformed. Instead, I shall begin by asking what it is we would like to know about the brain and its operations. A logical starting point for asking this question is the mind, for it is in the manifestations of our minds that we are immediately aware of the function of our brains.

In the course of reviewing how scientists have attempted to study the mind, I shall introduce a few notions about the different ways of looking at how the brain is put together and how its parts operate. This is important because there are always many different ways of answering questions about what the brain is doing. An answer about the reasons for behavior that satisfies a psychologist may not be satisfactory at all to a molecular biologist and vice versa.

Questions in brain research can be answered in many different terms. This will become apparent as we explore the dimensions of mind from the elusive notion of consciousness to the consequences of mental illness. Once we have completed this tour of the landscape of the mind, we shall be prepared to look inside the brain and ask whether anything there helps to illuminate the shadowy terrain of the mind.

Classifying brain cells
A computerized video display system helps researchers at the University of California, San Diego, to classify the billions of nerve cells in the human brain.

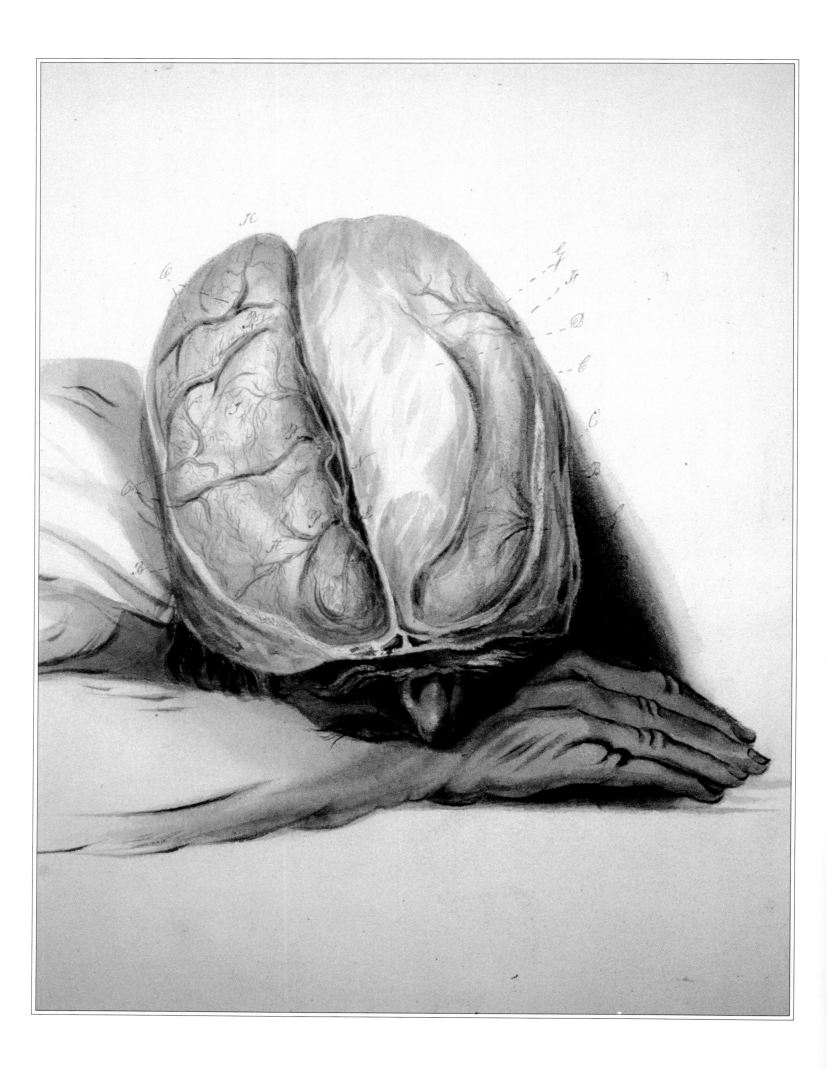

1
Consciousness: A Matter of Thought

How is consciousness defined and how might it be tied to what goes on inside the head? Originally the sole domain of the philosopher, consciousness has in recent years become a subject for brain researchers. As their studies have revealed more and more details about how the brain works, they have begun to ask how these details relate to the phenomena of consciousness. Is it a ghost they are chasing, or something concrete and measurable? Is it possible to say that the self lies in a particular part of the brain or is represented by some specific kind of electrical and chemical activity of its cells?

Not too many years ago schoolbooks showed pictures of the brain with a little cartoon man drawn on its surface, a homunculus who was supposed to represent the allocation of different parts of the brain to different parts of the body. By means of a homunculus it was possible to push out of the way the problem of consciousness and many other problems of brain function as well. Just make the homunculus responsible. "We are informed of our conscious awareness of ourselves and our thoughts by the homunculus," a knowing teacher could say. The teacher could also say that the homunculus enabled us to see the world around us by interpreting the patterns of light falling on our retinas. But the homunculus, squatting down beyond our eyeballs and peering out at the world through them, is not a very good scientific explanation these days. Even a child could ask if the homunculus has a homunculus and that homunculus another, and so on to infinity.

Some philosophers insist that trying to understand how the brain works, let alone its relation to the mind, is an exercise in futility. Their negative attitude is grounded in the belief that the brain or the mind cannot possibly understand itself. This is basically a matter of belief, although some try to dress it in fancier clothes by appealing to Gödel's incompleteness proof. This complex bit of mathematical logic demonstrates that there can be statements in a formal mathematical system whose truth cannot be demonstrated within that system. Powerful though it is in the realm of mathematics, Gödel's proof tells us very little about what the brain can or cannot do. But most brain research is not done by philosophers, and many of the people who do conduct such research go to work each morning with the expectation that some day we shall understand how the brain works and possibly even how brains are related to what we call consciousness.

Is consciousness a kind of door in the brain, wide open when we are alert, firmly shut when we have been knocked unconscious or anesthetized, and left slightly ajar letting in a bit of noise when we are lightly asleep? This is a view that relates consciousness to some sort of awareness of activity in our sense organs: our eyes, ears, or skin. Sights, sounds, and sensations of touch do activate the brain in certain ways, as we shall see later, but is that in itself adequate for consciousness? Is the self found in patterns of brain activity that somehow represent what is happening to us?

Some brain scientists, like Otto Creutzfeldt of the Max Planck Institute for Biophysical Chemistry in Göttingen, firmly hold a dualistic view of mind and brain. He maintains that the world of the mind, in which our consciousness exists in the form of symbols representing the physical world, differs from that physical world. This duality of symbols and physical reality leads Creutzfeldt to insist, "The knowledge of the brain mechanisms that lead to the consciousness that creates this world of symbols can explain neither consciousness nor the symbols in which it presents itself, since the brain mechanisms are not the symbols."[1] This line of argument ultimately leads Creutzfeldt to conclude, "There is, in fact, no way of defining or proving the existence of consciousness and mind in biochemical, biophysical, or anatomic terms."

The separation of mind and brain is even more dramatic in the work of Nobel laureate

Anatomy
This is a watercolor by Sir Charles Bell, a Scottish physiologist. It is taken from a series published in 1823 and entitled **The Anatomy of the Brain Explained in a Series of Plates.**

Consciousness: A Matter of Thought

John Eccles, famous for his basic work on how nerve cells operate. He proposes that the mind is wholly separate from the brain and acts upon the brain by making very subtle changes in the chemical signals that flow among brain cells. His ideas draw on the principles of quantum mechanics (the laws that apply to the interactions of particles on an atomic scale), but they attribute no physical properties whatever to the mind. The mind, in his view, exerts its effects by altering what are known as probability fields that influence the movement of messenger molecules between brain cells.

A very different way of looking at consciousness comes from Julian Jaynes, a psychologist at Princeton University. He draws together a bewildering array of evidence from split-brain research, archeology, ancient literature, and other sources to support his hypothesis that human consciousness arose some 3,000 years ago when the once independent hemispheres of the brain were forced by cataclysmic events to interact in new ways and to develop the ability to think in the sense that we know thinking today. Ancient humans, according to Jaynes, had no self-awareness.

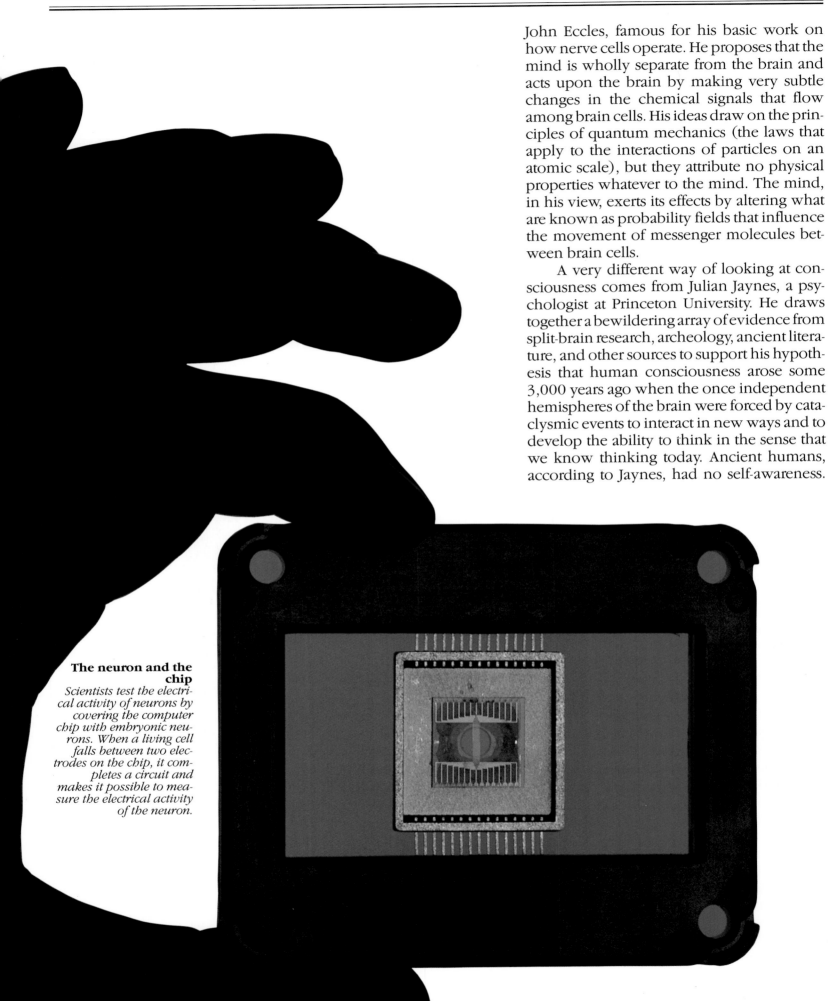

The neuron and the chip
Scientists test the electrical activity of neurons by covering the computer chip with embryonic neurons. When a living cell falls between two electrodes on the chip, it completes a circuit and makes it possible to measure the electrical activity of the neuron.

Rather, they operated under the influence of two autonomous but unconscious "minds," one in the right hemisphere and one in the left. He refers to this as the bicameral mind and calls the left hemisphere the "man part" and the right hemisphere the "god part."

In Jaynes's scheme of things, the man part of the mind functioned in much the same way as we can function today while driving a car or performing some other task so familiar that we can do it unconsciously. The god part of the mind was responsible for planning and guiding actions when novel circumstances occurred. It issued instructions to the man part through nerve fibers that connect the two hemispheres. The left hemisphere received these instructions as verbal instructions in the form of auditory hallucinations. Speeches of the gods in Homer's Iliad and other ancient literature supposedly reflect the nature of these hallucinations. The breakdown of the bicameral mind, which Jaynes claims is responsible for the origins of consciousness, was due to world turmoil in the second millennium B.C. Diverse cultures were forced together at that time, and this led to a loss of faith in the gods and the effortful introduction of thinking and self-awareness.

Jaynes's analysis of changes in ancient literature, which he attributes to the breakdown of the bicameral mind, supports his provocative hypothesis. The functions of the two hemispheres as we know them today are generally consistent with his argument, although some might dispute his interpretations. His arguments that auditory hallucinations and other aspects of schizophrenia may reflect a regression to a bicameral mind are plausible. Fascinating as it is, however, Jaynes's story depends at heart on psychological arguments and interpretive readings of literature. It is unsatisfying in that it tells us little about what to look for in the workings of the brain that might account for the troublesome phenomena of consciousness.

Have brain scientists found any physical clues to what consciousness is? One very ingenious researcher has done some experiments to explore the relationship between a person's brain activity and the conscious decision to make a movement. The results are perplexing and suggest that consciousness has the rather unusual role of gatekeeper. Benjamin Libet of the University of California, San Francisco, recorded electrical signals gene-

1. Homunculus
For the alchemists at the beginning of the Renaissance, homunculi were the children of the sun and the moon. Said to be produced by Paracelsus and Count Kuffstein, Chamberlain of Maria Teresa of Austria, homunculi symbolized the origins of alchemy.

2. "A representation of the interior"
From a Taoist treatise of the seventeenth century.

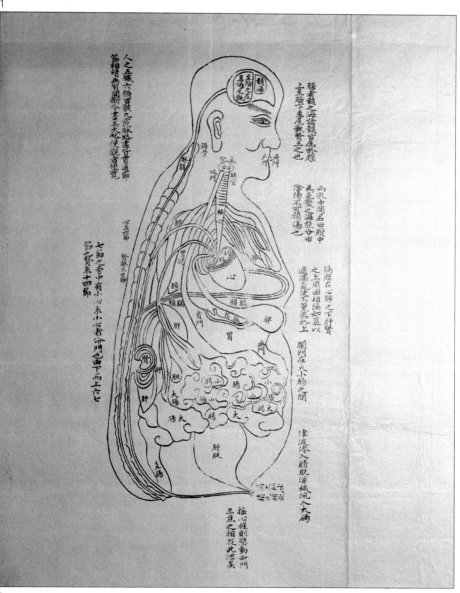

Consciousness: A Matter of Thought

1. The human brain according to René Descartes
"Imagine, for example, that the difference between the two figures M and N is the same as that between the brain of an awake man and that of a man asleep and dreaming." A view of the "machine of the mind" according to the author of Discourse on Method.

2. Sir John Carew Eccles
This famous Australian neurophysiologist, who received the Nobel Prize in Physiology or Medicine in 1963, is one of the staunchest defenders of mind-body dualism.

3. Measuring cerebral blood flow
This unusual instrument measures blood flow in the brain and makes it possible to compare the activities of different brain regions. Thanks to this new technology, scientists have acquired precious knowledge about normal and abnormal patterns of blood flow in the human brain.

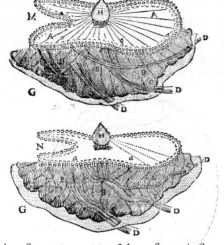

rated by the brains of his experimental subjects and looked particularly at a signal called the readiness potential that always appears just before a movement. Using special timing techniques, he found that the readiness potential begins about half a second before a subject begins to move a hand. This is expected, since brain activity must begin before the brain issues a command to the muscles. What is surprising, however, is that the subjects do not become aware of deciding to move until only about two tenths of a second before the movement begins, some three tenths of a second after the brain activity began.

There are many different ways of interpreting this unexpected result, but to Libet it says that the intention to act arises from brain activity that is not within our conscious awareness. In a sense, the brain initiates the impulse to act and the conscious self subsequently becomes aware of it. Libet also finds that his subjects are able to veto the impulse to act during the few tenths of a second after a subject becomes aware of it. In this sense, consciousness becomes a gatekeeper for intentions generated by the brain, letting through only those that somehow meet an individual's criteria.

Not everyone agrees with Libet's ideas about conscious will, but his experiments are part of a general change in the way scientists are thinking about the relationship between the mind and the brain. Old ideas about what the mind is and what it does are being scrapped as new knowledge about the brain emerges. We are clearly still a long way from understanding the old philosophical puzzles of consciousness. But researchers in neuroscience and cognitive science are becoming aware that they were not solving these puzzles because they were asking the wrong questions about the mind. Stephen Kosslyn of Harvard University characterizes the new view by saying, "The mind is what the brain does."

With this assertion, I shall set aside the question of consciousness for now, picking it up again when we have examined in some detail what we know about what the brain does. As Kosslyn optimistically states, "For the first time we have the concepts to guide us in looking at how the brain makes the mind, and the tools that allow us to look." This means starting over again in the quest for mind and asking new questions that relate to what we know about how the brain works. Toward the end of the book I shall return to this subject to see how knowledge about the brain helps to formulate new questions.

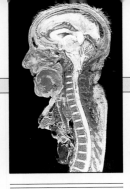

Brain Landscapes of Consciousness, Language, and Dementia

by David H. Ingvar, M.D., Ph.D., University Hospital, Lund, Sweden

The brain imaging techniques that arose in the early sixties enabled for the first time topographical displays of brain activity, not only, as previously, brain anatomy. It thus became possible to detect changes of brain function, mainly changes in the activity of the neurons of the cerebral cortex (which are related to the arrival of sensory impulses and the execution of voluntary motor acts), as well as mental activity in normal and diseased brains.

The first techniques of functional brain imaging were based upon the xenon-133 technique developed by Lassen and Ingvar. It principally uses batteries of detectors that record the arrival and disappearance of the freely diffusable isotope xenon-133 at the side of the head. The rate at which the isotope disappears (its clearance curve) in a given brain region is a function of the blood flow. Since the flow is normally determined by the work (or energy metabolism) of the neurons, the level of the flow is an index of the activity of the neurons in the regions measured.

During the seventies and eighties a number of new brain imaging techniques were developed. These are all based upon the use of various radioisotopes that follow the blood flow or the metabolism of oxygen, sugar (glucose), proteins, or most recently, neurotransmitters. Especially noteworthy is the brain imaging technique of positron emission tomography (PET), by which one can obtain "slices" of the brain at different levels showing metabolic activity. The xenon-133 technique is still widely used to study normal and abnormal mental events in the human brain. Here three examples are given of functional-flow landscapes in the cerebral cortex. In figures 1, 2, and 4 the brain is seen from the left side with the frontal lobes to the left and the occiput to the right. In figure 3 the brain is seen from the top, showing the activity in both hemispheres.

Early in our studies my colleagues and I observed that the activity distribution in the cerebral cortex is not uniform in the resting state, i.e., when a subject lies supine, awake, undisturbed, not spoken to, not performing any movements, nor solving problems. The cortical flow activity is significantly higher in cortical regions anterior to a line that approximately follows the central (rolandic) fissure and the anterior part of the lateral (sylvian) fissure. Grossly, these regions correspond to the efferent (output) parts of the motor cortex in the precentral gyrus, as well as the more frontal, higher control centers of complex movements, behavior, and cognition. Postcentrally the activity is lower. Located here are regions that have an afferent (input) function related to the primary, higher-order projection areas for cutaneous sensation, vision, hearing, etc.

The hyperfrontal-flow landscape in the resting human brain disappears under general anesthesia and severe organic dementia, and is also altered in some forms of mental disorders. In addition, there is some evidence that it disappears in sleep.

The relatively high resting activity of the frontal regions (on both sides) increases during complex motor performance, including speech (see figure 2), as well as during sensory perception and cognitive activity such as that of problem solving. Such observations support the hypothesis that the resting hyperfrontal pattern represents a simulation of behavior. A basic function of the frontal lobes thus appears to be the production of sequential neuronal pro-

The resting landscape of conscious awareness

1. The landscape of resting consciousness
Measurements with xenon-133 of regional cerebral blood flow in six resting, undisturbed, normal subjects. The mean flow was 48 ml/100 g brain/minute. Regional deviations from the mean are shown in relative terms (see scale on the right side). Note the high blood flows in frontal regions and the low blood flows in posterior parts of the cortex. This is the "hyperfrontal" pattern of resting consciousness. (From D. H. Ingvar, *Acta Neurol. Scand.* 60 [1979]: 12-25)

2. Reading
In the top image the subject reads aloud. Note that activation is seen in several regions, such as visual, auditory, and motor-mouth regions (larynx, lips, eyes, etc.). Speech centers in the upper and lower frontal areas are also activated. In the bottom image, when the subject reads silently, the mouth-larynx area and the temporal auditory area are not activated. Relative flow values are given by the percentual scale to the right. (From N. A. Lassen, D. H. Ingvar, and E. Skinhöj, *Scientific American,* October 1978)

3. Speaking and singing
Bilateral, xenon-133 brain-blood-flow measurements in 15 normal subjects. Each pair of images show brains at rest (upper left pair), during speech (upper right pair), and while humming a children's song. The color scale at the right side shows the actual mean regional flows. Frontal regions are at the top of each pair, where mean hemisphere flows are also given. Note that speech activates large, mainly anterior-frontal areas and that bilateral speech centers also show up. Humming gives a similar pattern without activating centers that control spoken words. (From E. Ryding, B. Brådvik, and D. H. Ingvar, *Brain* 110 [1987]: 1345-1358)

4. Dementia
The upper picture shows the normal mean resting landscape of conscious awareness in six subjects (the same as figure 1). The lower shows the average landscape in ten demented patients. Relative values are given, as in figure 1. Note that the mean flow in the demented group is reduced to 35 ml 100 g brain/minute (48 in the normals). Also note that there are relative distribution abnormalities in dementia, with low flows in frontal regions and especially in the temporal and parietal regions. Demented patients of this type show memory disturbances, disorientation, emotional abnormalities, indifference, and other symptoms. (From D. H. Ingvar and N. A. Lassen, in *Brain Function in Old Age,* ed. A. Hoffmeister [Springer, 1979], pp. 268-277.)

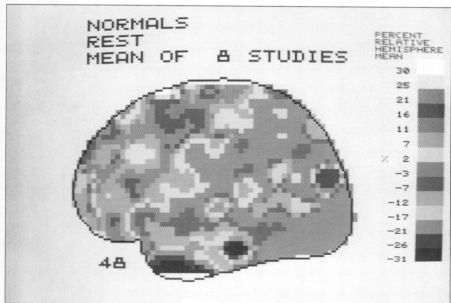

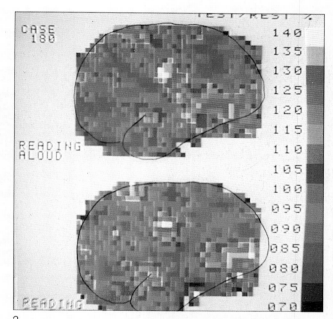

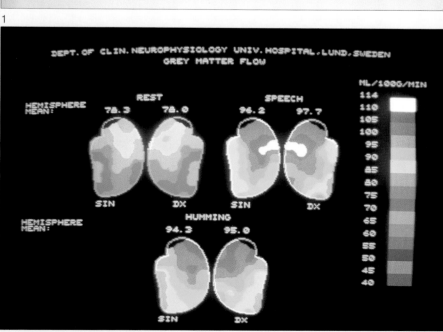

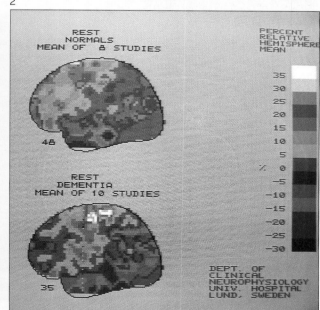

grams that underlie conscious awareness and can be mobilized for various types of motor behavior, speech, and cognitive activity.

The landscape of language

In figure 2 regional blood flow changes are seen in the cerebral cortex when a subject reads a text aloud (top) and when he reads the same text silently (bottom). As is seen, several cortical centers are activated. Visual centers are activated in the occipital region when one reads a text aloud and so are adjacent language centers necessary to interpret the printed symbols. Also activated are the mouth, tongue, and larynx regions in the lower rolandic region, which controls the speech production of the reader. In addition, the auditory feedback from the voice of the reader (who listens to what he says) activates hearing centers in the upper temporal region. Language centers in the lower and upper frontal lobes also show a high flow, as do the motor centers controlling eye movements, which follow the text as it is read. When the same text is read silently the auditory and mouth regions are clearly no longer activated. Otherwise, silent reading requires activation of the same centers in the cortex as reading aloud.

This example demonstrates how the xenon-133 clearance technique can be used to study rapid changes in the activity of cortical nerve cells. Each of the studies takes about two minutes to perform. The resolution is limited to about 1 to 1.5 cm, but even so, distinct regional changes in cerebral activity pertaining to reading are depicted in detail.

Landscapes of speech and singing

In figure 3 the xenon-133 technique has been used with bilateral detectors, thirty on each side, and the picture shows average flow changes in fifteen subjects without brain disease (see the scale to the right). All diagrams show the brain seen from above with the two hemispheres folded out to the sides and the frontal lobes at the top.

The three diagrams show the resting state, speech (saying the days of the week), and humming the tune of a children's song. The resting state (top left) shows the normal hyperfrontal landscape of the resting state (see figure 1). Speech augments the flow of the frontal regions, and there are two centers in both hemispheres that show the highest flows (top right). They represent centers that control the movements of the mouth, tongue, lips, larynx, etc., necessary for speech (see figure 2). The frontal activation is most likely related to the sequential programming of speech performance, which, though of a simple nature, requires a detailed control of the words produced. In the bottom image when the subjects were humming, frontal activation is also evident, but the speech centers do not show up. Humming a melody also requires detailed frontal-lobe control of the voice, but there is no word production to be controlled.

This example shows that the whole brain is active during speech and humming and that it is not only the left hemisphere that controls speech, as had been presumed on the basis of the classical observations of Broca and Wernicke in the last century (approximately 1860-1875). In fact, detailed analysis of the results shown in figure 3 demonstrates that during speech and humming the right hemisphere works a little bit more than the left hemisphere.

The landscape of dementia

In figure 4 two average landscapes are shown from six normals and from two patients affected with organic dementia of the Alzheimer type, in which there is a loss of intellectual function due to degeneration of the nerve cells of the cerebral cortex. Two facts are evident. The landscape of the normal brain at rest is of the hyperfrontal type as described in figure 1. Note also the high mean flow (48 ml blood/100 g brain/minute). In the demented patients the flow level is reduced to 35 ml/100 g/min, and there are also abnormally low flows in the landscape. Both in the frontal lobes and especially in the parietal and temporal regions, the flow is very reduced. Note that in both figures relative levels are given in relation to the mean flow. This means that the frontal blood flow in the normal achieves a value of about 70-80 ml/100 g/min, while the frontal flow in many regions in the demented cases is only 20-30 ml/100 g/min. Thus, the brain activity in dementia is accompanied by a reduction of flow in the frontal and parietal regions to less than half of the normal.

In demented patients the defects in the cortical flow landscape correlate with intellectual defects. The overall flow reduction is proportional to the overall intellectual deficit. Regionally it has been found that if the flow is low in the temporal regions, this is accompanied by memory defects. Low flow in parietal regions is found in patients who have lost their orientation and ability to handle tools (spatial agnosia and apraxia). A low frontal flow is found in patients who are inactive and content yet have no idea of the extremely tragic future of their state.

Figure 4 thus gives an example of how brain imaging techniques using xenon-133 are used to measure defects in cerebral functions, in this case due to organic dementia. Measurements of this type may also be used to test the effects of various forms of therapy, including the administration of drugs.

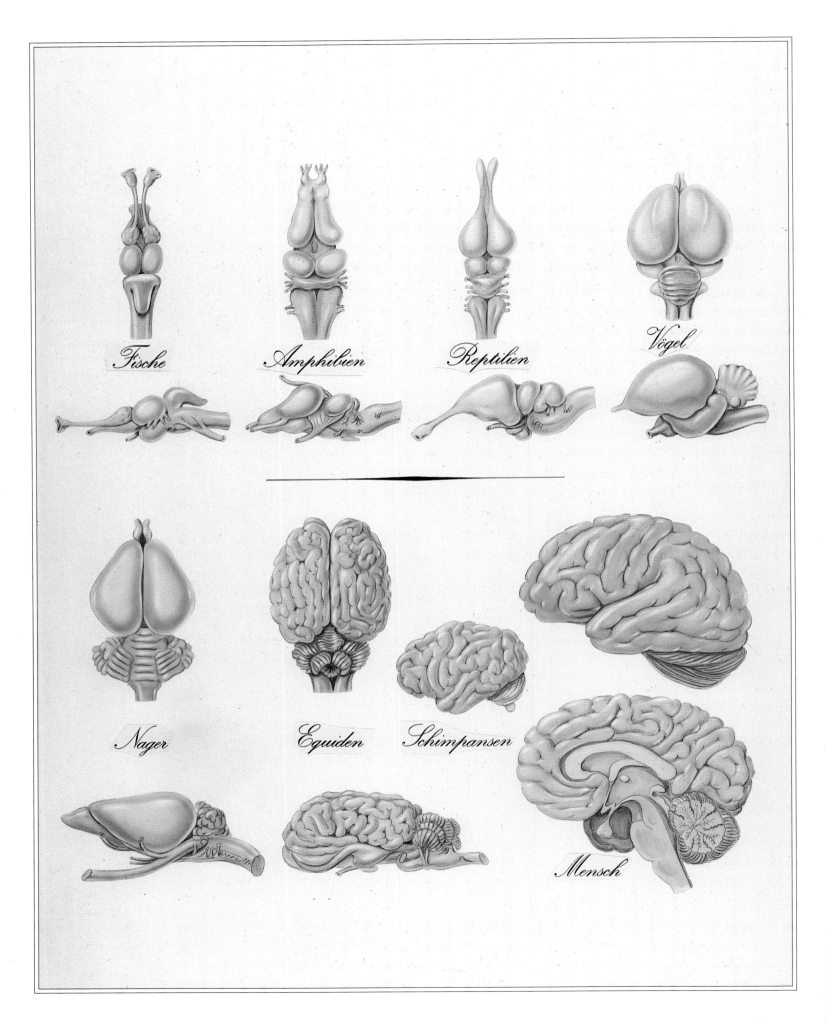

2
Brain Power:
The Search for Intelligence

If at least for now consciousness proves too elusive a subject to understand in terms of brain function, perhaps some more concrete aspect of the mind might be identifiable in terms of brain function. Once again taking a cue from ancient philosophers, let us look at the processes of rational thought, the kind of thought associated with playing chess or doing mathematics. These are processes that some consider to be the highest accomplishments of the human brain. Is there any way to determine how the brain carries out these very precise steps of logical analysis? Will answering the question tell us anything about the important capabilities of the human brain?

Intelligence is a concept that is at once both vague and precise. Everyone has the ability to judge whether or not someone is intelligent or very intelligent or just a bit above average. We are well aware that intelligence testing is a widespread practice used to screen applicants for jobs or students for placement in schools. But what do intelligence tests measure: one or more specific capabilities of the brain? We all know that many people who do well on intelligence tests seem unable to perform well in their everyday lives. The opposite is true as well. Some observers think that intelligence tests provide a measure of how well you will succeed in life, while others think that all intelligence tests do is tell how good you are at taking intelligence tests.

It is perfectly possible to make a career of studying human intelligence without even considering the brain. The earliest intelligence tests were largely empirical, created by finding out what kinds of questions people who are considered intelligent are good at answering. As the tests became more refined, they became more specific about the kinds of intelligence they purported to measure, sorting out such factors as verbal intelligence and mathematical intelligence. Some psychologists developed elaborate systems of mathematical analysis, attempting to sort out different kinds of intelligence, while others maintained that there is only one fundamental kind of intelligence somehow related to mental energy. Few paid attention to what the brain does to manifest mental ability.

One recent theory of intelligence that does pay attention to brain research is a theory of multiple intelligences developed by Harvard psychologist Howard Gardner. He proposes that there are at least seven distinct categories of intelligence: linguistic, logicomathematical, musical, spatial, bodily movement, interpersonal, and intrapersonal. Gardner derives these intelligences from many types of observations, among them observations of the effects of brain damage on people's ability to perform certain mental tasks requiring some intelligence, such as reading stories or drawing pictures. Gardner draws largely on the clinical evidence that brain damage can affect certain of these abilities, while leaving others relatively untouched. Some researchers question the independence of these intelligences, since some of these abilities, like the logicomathematical and the spatial abilities, are difficult to test independently. Nonetheless, Gardner's ideas do provide some links between intelligence and brain function, even though they are not very specific about what that brain function might be.

It is extremely difficult to observe the human brain in action, that is, to look directly at what individual cells in the brain are doing. But because much of the action of those cells involves electrical signals, it is possible to observe the overall patterns of electrical activity generated by the brain. Small pieces of metal placed on the scalp can tap into the changing electrical fields produced by the collective activity of many brain cells. These electrical signals can be recorded as undulating lines on paper that are called electroencephalograms. If you flash a bright light or make short bursts of sound, the brain produces characteristic electrical responses known as evoked potentials. These evoked potentials provide one way to look at what happens to

The evolution of the brain
The evolution of the vertebrate brain is dramatically illustrated by these samples that span hundreds of millions of years. From left to right and top to bottom are the brains of a fish, amphibian, reptile, bird, rat, horse, chimpanzee, and human.

brain activity while human beings perform various mental tasks, and they are important tools in several aspects of contemporary brain research.

Since evoked potentials reflect something about what the brain is doing, can they tell us anything about intelligence? Jane M. Flinn and Arthur D. Kirsch of the George Washington University, working with Edward A. Flinn of NASA, studied a group of 12-year-old girls whose IQs ranged from 69 to 137. They recorded from the girls' scalps electrical signals that occurred in response to flashes of light and analyzed the signals to determine the relative amounts of low-frequency and high-frequency energy. The results indicate that girls with high IQ tend to have more high-frequency energy in their electrical responses than low-IQ girls, while the latter have more low-frequency energy than the girls with higher test scores. This kind of analysis provides a fascinating but indirect glimpse of how brain processes associated with intelligence might be probed, but it tells us very little about how intelligence as measured by IQ tests relates to what the brain actually does.

Another way of looking for the elusive quality of intelligence in the brain is to look at the brain's structure. Like well-developed muscles, well-developed intelligence might be visible in some aspect of the structure of the brain. Mark R. Rosenzweig and his colleagues at the University of California, Berkeley, compared the brains of laboratory rats that had been raised in enriched environments with those of animals raised in standard cages. (The enriched environments contained a small community of animals, along with many toys and opportunities for exploration.) They found many differences between the two groups of brains, particularly in the cortex, the thin layer of gray matter that presumably mediates what are referred to as higher mental activities.

Animals from the enriched environment proved to have a thicker layer of gray matter than that found in deprived animals. The cells in the cortex also made more connections with other cells in the animals from the enriched environment. There is, however, a serious problem in interpreting these results. While considering brain development in a later section, we shall find that early in life there are more brain cells and brain-cell connections than the adult brain needs. So it is possible to argue that brains with less luxuriant growth may actually function better. We have yet to learn which view is more accurate.

Interesting as these results are, they have more to do with experience than with what we might consider to be native or inborn intelligence. Over the years there have been many attempts to make postmortem comparisons of the brains of highly intelligent persons with those of less gifted individuals. These studies have not produced any results worthy of mention, in part because changes that occur with aging may well mask any differences that relate to intelligence, particularly since no one has really known what to look for.

One recent study of the brain of a notably intelligent individual has attracted considerable attention, perhaps more for its strange history than for the scientific insights it provided. This was the examination of Albert Einstein's brain by Marian C. Diamond and her colleagues at the University of California, Berkeley and Los Angeles. The circumstances surrounding the removal and preservation of Einstein's brain following his death in 1955 have been a subject for speculation for some years. What is known is that Einstein's brain came to light in recent years in a doctor's office in a small town in Missouri. Given an opportunity to examine the brain, Diamond and her colleagues were then faced with the problem of what to look for, what questions to ask of Einstein's brain.

The decision they made was based partly on Einstein's own reports of how he solved problems and partly on what Diamond and others, including Rosenzweig, had found in the brains of rats raised in enriched environments. Einstein was on record as saying that his productive thought relied on associations between signs and images. The rats from enriched environments showed a relative increase in the numbers of glia cells (cells that play a poorly understood supporting role for nerve cells) compared with the nerve cells in the cortex. Taken together, these two factors led Diamond to look at two regions of Einstein's brain considered to be involved in intellectual neural functions. They do not receive any input directly from any of the sensory systems, but rather collect their input signals from widely distributed regions of the cortex. Because of this, they are referred to as association cortex. The question Diamond asked was, How do the ratios of nerve cells to glia cells in regions of Einstein's association cortex compare with the ratios found in normal subjects?

While we are not yet ready for a description of the anatomy of the brain, it is appropriate here to explain briefly what glia cells are and what they do. The most important thing to say is that we are not really sure what their role in brain function might be. The name "glia" comes from the Greek word for glue, and that was the rather dull role they were assigned for many years, just the material that holds together the far more important nerve cells. There are about ten times more glia cells than nerve cells in the brain, however, and some brain researchers have suggested that it might really be the glia cells that are doing such important things as storing memories. Evidence now available suggests that the real role of glia probably has something to do with maintaining the proper chemical environment for nerve cells, but that is still just an educated guess. These parenthetical comments raise at least one cautionary note about the search for a physical manifestation in Einstein's brain.

The one measure that proved to be significantly different from normal in Einstein's brain was the neuron to glia ratio in the area known as the inferior parietal lobule. It is a region of cortex implicated in the association of information from cortex concerned with vision, audition, and touch. Individuals who suffer damage to this region often appear to have problems with complex thinking as a consequence. In Einstein's brain, the neuron to glia ratio in this region of his left hemisphere was 1.12, while the average for the other brains examined was 1.936. This result may or may not give some support to the notion that brain structure is somehow related to intelligence, or it may show the inadequacies of statistics in such difficult cases. Still, the goals of finding out how brain function relates to intelligence and using this knowledge to take the greatest possible advantage of available intelligence remain distant objectives.

We assume that intelligence evolved along with the brain, so we might expect to find some clues to the nature of intelligence by trying to retrace the evolution of *Homo sapiens* and attempting to reconstruct how the intellectual capabilities of such earlier forms as *Homo erectus* developed while the physical characteristics of their brains changed.

1. Albert Einstein
Nobel Prize winner in physics and talented violinist, Albert Einstein was considered one of the most intelligent men in the world. Marian Diamond at the University of California, Berkeley, has found some unusual features in the structure of his brain.

2. A lesson in mathematics
In Japan education in all fields of knowledge begins early to prepare the child for intense competition in life.

3. Aristotle
The Greek philosopher Aristotle, born in the fourth century B.C., was the father of logic and the prince of philosophers.

Brain Power: The Search for Intelligence

The most intelligent of birds
A macaw (Anodorhynchus hyacinthinus) was trained by Animal Behavior Enterprises to select puzzle pieces that fit the cutouts on the board. This behavior corresponds to performances tested by certain intelligence tests and tests for mental handicaps in children. It involves visual discrimination of color and shape as well as the motor performance required to put the piece in place. The macaws, along with other members of the corvine family (crows, magpies, jays, etc.), is regarded as the most intelligent of birds. These birds can learn to perform such complex sequences of behavior as that shown here.

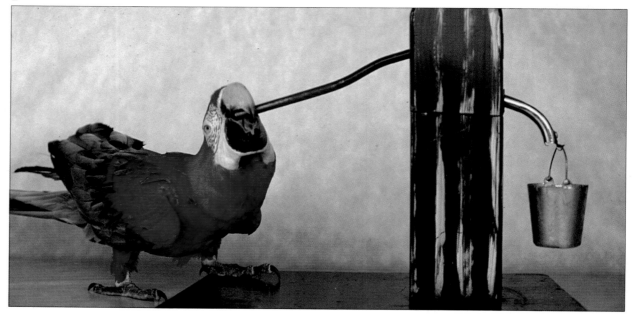

There is no guarantee, of course, that the intellectual changes are going to be the direct result of the changes in the brain. And furthermore, since our illiterate ancestors left very few physical traces of their mental activities, discovering the intellectual changes is a very difficult challenge.

One physical trace that Leon Festinger of the New School for Social Research thought to be revealing is the form of stone tools made by early humans. These can be found in abundance in archaeological sites throughout the world. *Homo erectus,* who appeared on earth about 1.5 million years ago, made primitive stone tools that remained monotonously unchanged for over one million years until the species disappeared a few hundred thousand years ago. With the subsequent emergence of our immediate ancestor, Neanderthal man, things began to change rapidly. Neanderthal man could produce tools of greater sophistication, and what we consider greater aesthetic appeal, than anything that came before. But what made the difference?

There are two bits of suggestive evidence about the sudden spurt of technological development that occurred upon the arrival of Neanderthal man. One is suggested by the work of French archaeologist Jacques Tixier, who has learned the difficult skill of making tools in the Neanderthal manner. The complexity of making these relatively advanced tools suggests that unlike earlier crude implements, their production could not be efficiently taught to others without the use of language to communicate. The implications are profound, if indeed the development of advanced tool-making skills marks the emergence of the advanced cognitive capability we know as language.

If we seek to trace the evolution of the brain back farther into the mists of prehistory than the time of primitive man, the task seems impossible, because in fossils all traces of the microscopic structure of the brain have been erased. Nonetheless, we do have a living resource in the brains of animals that are, in one way or another, distant relatives of one another. By comparing the brains of creatures that have different evolutionary ages, brain scientists like Sven O.E. Ebbesson of the Louisiana State University at Shreveport hope to learn some clues about the development of brain power as animals became capable of mastering an ever-increasing repertoire of behavioral skills. His studies have led him to develop what he calls the "parcellation theory" of the evolution of the brain.

The basic idea of parcellation is rather simple and straightforward, but mustering the evidence to support it is a formidable challenge. We can take only a brief glimpse at it here; the ideas will become clearer after the reader has absorbed later discussions of brain anatomy and the formation of the brain early in life. The notion is that as brains evolve, different parts of the brain become more isolated from one another and more specialized in their functions.

In primitive brains we find different parts of the brain connected together with many nerve fibers. These many different parts are all involved at the same time in performing such functions as the perception of touch and the control of movement. In higher mammals such as monkeys we find that many connections between different brain regions are lost; the control of movement and feeling are distinct functions of quite separate brain regions. Much of the evidence to support this parcellation hypothesis comes from the observation that during the formation of more advanced brains, many brain regions are connected together at an early stage, and such connections disappear as the brain matures. This is discussed in more detail in the section on brain development.

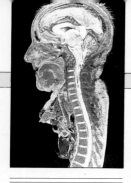

The Brain and Evolution

by Yves Coppens, professor at the Collège de France, director at the Musée de l'Homme

The Hominidae family appeared around 7 or 8 million years ago. The first known branch, *Pre-Australopithecus,* made its appearance in Tanzania, Kenya, and Ethiopia over 6 million years ago. It was followed by *Australopithecus,* which appeared in West Africa and South Africa between 3.5 and 1 million years ago. The genus *Homo* first appeared in East Africa with *Homo habilis* (between 4 and 1.6 million years ago) and *Homo erectus* (approximately 1.6 million years ago). The third branch from which modern man has descended, *Homo sapiens,* appeared at least 100,000 years ago.

The only means available to paleontologists for understanding the evolution of the brain over these millions of years is the cavity of the skull and a few indirect clues (tools, traces of rituals, etc.). The paleontologist measures the volume of the cavity and makes a casting of its surface to observe the obvious traces left by the meningeal vessels.

Cranial volume

To begin with we should specify that by volume (calculated by a number of different methods) we mean the volume of the whole endocranial cavity and not simply that of the brain. The percentage occupied by the brain in human fossils with respect to the rest (meninges, cerebrospinal fluid, etc.) cannot be calculated with any great accuracy, but this combined volume is nonetheless a very useful indication. Over the millennia the volume of the brain definitely increased, although the process was far from regular. In *Homo erectus,* the increase is dramatic, with a parallel development in cultural activity, such as it can be reconstituted.

But the volume or weight of the brain is not a particularly useful indication in itself. It must be compared with other parameters like size or volume of the body. At one point size seemed an important factor: in the brontosaurus and the great dinosaurs the brain represents 1/100,000th of total body weight, in the whale 1/10,000th, in the elephant 1/600th and in man 1/45th. Unfortunately for this theory, the brain/body ratio in mice is 1:40 and in marmosets 1:25! Recently Roland Bauchot of the University of Paris VII and Heinz Stephan of the Max Planck Institute have formulated a theory of brain-weight ratio according to which the ratio of brain weight to total weight is higher in diurnal than in nocturnal species, in ground-dwelling as opposed to tree-dwelling species, and in bipeds compared to quadrupeds.

Castings and meningeal vessels

Castings provide a wealth of valuable information, such as that over the course of evolution the associative (lateral) zones increased at the expense of the visual zones (the occipital lobe) and that the height of the brain and the number of convolutions also increased. Ralph Holloway at Columbia University has developed stereoscopic methods for marking out the surface and quantifying changes in shape. While there is some truth in Leroi-Gourhan's remark that man must accept that his history began with his feet, researchers have nonetheless tended to neglect the early transformation of brain structures. Judging from the external morphology of the castings, there were already great differences between the Hominidae family and that of the great apes.

Mental activity requires energy in the form of oxygen carried by the blood, and increased brain size implies better irrigation of the brain. The development of the meningeal vessels, the only system accessible to the paleontologists, since it is imprinted on the inside surface of the brainpan, is spectacular: two vessels in the oldest australopithecine, three in more evolved australopithecines, a fairly dense network from *Homo habilis* onwards, and a complex and very dense network in the first *Homo sapiens*.

The appearance of technical intelligence

Australopithecines struck a stone to obtain a splinter, then shaped the splinter by striking it. From *Homo habilis* onwards, the paleontologist can draw up quite an extensive list of tools. Different types of tools were made and tried out, and those most suitable for a particular use were retained and then reproduced, which implies the existence of a learning process and the ability to take on long-term projects. The first examples of ritual practices seem to have appeared around one million years ago, in Java and China, where paleontologists have unearthed skulls sectioned down to the brainpan. The accuracy allowed by electron microscopes shows that the front and base of the skull were sectioned using flints. The ornament appeared much later with *Homo sapiens,* and represented a preliminary approach to the world of signs, symbols, and abstract thought.

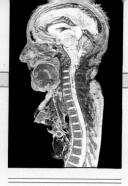

Lemurs, Ancestors of Our Ancestors

by Véronique Barre and
Jean-Jacques Petter,
Musée d'Histoire
Naturelle

The order of primates is made up of two suborders, prosimians and simians. The prosimians are considered more primitive, since the original form appeared before the first simian (apes and monkeys). All primates are adapted to tree dwelling, but the prosimians differ from the simians in the many primitive characteristics they possess, especially relatively small brain size and more lateralized vision.

Different prosimian species are found in the tropical and equatorial zones of Africa and Asia, but lemurs are found only in Madagascar and constitute the most strikingly original group of prosimians. Lemurs vary considerably in size; they range from the small dwarf lemur to the indris lemur. Most lemurs have long tails and enormous eyes. The species diversified greatly in Madagascar both in terms of contemporary forms as well as now-extinct forms. Lemurs are of particular interest because of this diversity of species. In addition, the primitive characteristics of lemurs serve as a reference and can contribute to our knowledge of some of the adaptations our own species has made and to a better understanding of primate evolution. Scientists believe that all the different subspecies of Madagascan lemurs developed in isolation in Madagascar itself.

One of the major characteristics of the primate order is the tendency toward increase in relative volume of the brain and the development of a cerebral cortex. The primate brain may have evolved in parallel with the development of vision, since development of the cerebral cortex occurs simultaneously with improvements in the organs of sight. It is obviously essential that tree-dwelling animals have very accurate vision to move about in trees and calculate leaps. The eye became the dominant sensory organ, and the brain evolved alongside. When we look at primate evolution and in particular at the evolution of lemurs, we note that the animals tended gradually to become diurnal rather than nocturnal and the eyes moved to a more frontal position to provide three-dimensional vision. The sifaka lemur can perform very accurate leaps of up to ten meters. It is obviously essential for the animal's survival that it not miss its target. This kind of accuracy demands the ability to analyze, memorize, and compare visual images, which in turn demands a corresponding development of the brain. Because of its small size, the dwarf lemur has less to fear from a fall, but it is hardly surprising that its brain is more highly developed than that of a small, ground-dwelling insectivore.

Along with the development of the brain and vision, enlargement also tended to occur. Small mammals are very vulnerable to predators, and in the course of evolution, natural selection will eliminate some of these species. One way of overcoming this problem was to grow larger. Any species whose descendants tend to do this will be favored. For tree-dwelling species like the primates, however, the most significant consequence of this evolution was adaptation to dwelling on the ground. A small species like the dwarf lemur can move amid twigs and creepers and on small branches because of its light weight. A slightly larger species such as the sportive lemur has less freedom of movement, since it cannot move around on the ends of branches. Where size increase is even greater, as in the sifaka lemur, the animal can use only the tree trunk as a support. For very large species, leaping becomes increasingly dangerous and exhausting: the indris lemurs tire easily and tend to descend toward and even onto the ground, leaping from the base of one tree trunk to another. On the ground their bodies remain vertical. It is easy to see how the evolutionary process would continue.

The lemur changed from an almost solitary, nocturnal animal into a diurnal animal living in family groups. Other factors are also involved in changing social behavior, but we must stress the importance of vision and communication in the improvement of interindividual relations: social life requires that aggressiveness be controlled and that there be a whole network of behavior, which can only be regulated by a very complex brain.

Many lemurs become active only in the evening or after nightfall and have excellent night vision. Night vision differs from daylight vision because of the lack of distant markers: the animal moves from one visual sphere to another and locates objects in very dim light. Three remarkable adaptations account for this ability. The first is the existence of very large globular eyes, placed slightly to the side. The animal can thus capture large amounts of light and see upward and sideways. The second is the presence of a tapetum or iridescent choroid membrane, a layer of light-reflecting cells in the back of the eye behind the retina. This membrane reflects back to the retina the light that strikes it so that the double stimulation of photoreceptor elements in the retina results in an increase in sensitivity. This reflecting capacity gives the eye a golden color when light shines on it. It tends to disappear in diurnal species, since it is an adaptation to nocturnal life. In the catta and indris lemurs it tends to be infiltrated with black pigment, so that if a light shines onto the eye, dark- and light-colored zones appear alternately. In simians this membrane is completely covered with black pigment. The presence of some fluorescent cells in the tapetum constitutes a third advantage for night vision. By receiving violet and ultraviolet light, which human beings are unable to perceive, these cells transform ultraviolet light into a greenish-blue light, which stimulates the cells of the retina to their maximum sensitivity.

Three-dimensional vision, which depends on binocular vision, is an essential faculty of primates. It does exist in the dwarf lemur, although the fields of vision of the two eyes intersect less than in diurnal species, even if the total visual field is slightly greater. Frontalization of the eyes is especially notable in the sifaka and indris lemurs, which move about from trunk to trunk. As the eyes move back into the head and become gradually more frontal in diurnal species, the field of vision is gradually limited in exchange for improved focal vision. As a diurnal species becomes larger and correspondingly stronger, it is less vulnerable to outside attack, especially as social bonds become stronger. The advantages linked to frontalization therefore tend to compensate for any disadvantages.

1. Woolly lemur
This small lemur, the only nocturnal representative of the Indriidae family, lives in family groups. It is a leaf eater and, along with the dwarf lemurs, is one of the most primitive lemurs. Note the light reflecting from the eyes.

2. *Microcebus murinus*
One of the most primitive and smallest of the lemurs, the *Microcebus murinus* eats anything, including resins, fruits, and insects. It spends the day sleeping in tree trunks or nests. *Microcebus murinus* most likely crossed the sea from Africa to Madagascar on tree trunks. Except for humans, and lemurs no primates live on the island.

3. Tapetum lucidum
This light-reflecting membrane lies behind the retina of the eyes of lemurs and allows light to pass through the retina twice, thereby improving night vision. It accounts for the reflected light seen in figure 1. It also occurs in cats, where it produces the same effect. Later in evolution in the simian eye the tapetum became totally invaded with black pigment.

4. Aye-aye
Very rare today, the aye-aye is an excellent example of morphological adaptation. It can pierce a coconut with its teeth and extract the meat with the aid of its highly specialized third finger, which it uses here to scratch itself. It also plunges the finger into the bark of trees to extract larvae. Its powerful grip permits it to climb and to adopt acrobatic postures to capture prey.

5-6. *Propithecus*
Propithecus is a diurnal member of the Indriidae family. It is a leaf eater and lives in family groups. It marks its territory by rubbing against tree trunks to leave a scent. Like the woolly lemur, it can stand upright. It can also perform impressive soaring leaps.

3
The Minds of Animals: Behaviorism and Ethology

Many scientific efforts aimed at understanding how the brain works involve the study of animal brains. And much of what we know about how the human brain works derives from these studies of animal brains. It is also true that many of our ideas about human behavior derive from studies of animal psychology. Yet over the years there has been sharp controversy among workers in these various fields, controversy not only over the previously discussed connections between mind and brain but also over whether animals have minds, whether states of mind or even events in the brain are important in behavior, whether there is such a thing as innate behavior, and a long list of related questions. Many are yet to be answered. We shall examine some of these questions here in the context of historical developments. Once again it will become clear that getting useful answers to research questions critically depends on knowing what questions to ask.

There are deep historical roots for many aspects of the study of mind, brain, and behavior. One major branch of psychology derives from the thoughts of Aristotle, who proposed that the human mind is blank at birth and depends wholly for its functioning on the acquisition of knowledge through experience. Some centuries later René Descartes proposed quite the opposite, that much of human knowledge comes prepackaged in the brain. The pendulum swung forcefully back in Aristotle's direction in the seventeenth century, when British philosopher John Locke established the empiricist school, asserting that all knowledge is based on learning. Locke thus set the stage for the emergence of a psychology centered on the study of the processes of learning. Throughout the seventeenth and eighteenth centuries, this psychology focused on the human mind. Scholars of learning considered a mind to be essential to the learning process, and since only humans were considered to have minds, there supposedly could be no such thing as animal learning.

The next major advance came when the work of Charles Darwin placed humans, over much objection, firmly in the animal kingdom. Some years later in America, a new breed of psychologists developed the idea that the study of learning in animals could be carried out without any reference to ideas or mental states. John B. Watson first formulated the basic tenets of a new field of behaviorism, a psychology that emphasized the objective observation of behavior without reference to any unobservable mental acts. As we can see today, rejecting insight into one's own mental activity as a valid source of data is justified, although not necessarily for the same reason it was thrown out in those early years. It is becoming evident now that introspection into how we solve a problem or recognize a face or perform some other intelligent act does not necessarily give a good picture about what the brain is actually doing. Sometimes it can even be quite misleading. Nonetheless, any adequate theory about what the brain is doing to carry out such acts must explain why we think they happen as we do.

Behaviorism reached its extreme in the work of B. F. Skinner. His position is one that scrupulously avoids any speculation about unobservable events that may be taking place in the brain. He denies any need for the mental constructs of consciousness or the physiological constructs of brain theory. His psychology is a description of behavior solely in terms of behavior. Much Skinnerian research concerns the factors that lead to changes in behavior. The terminology is complex and often depends on subtle distinctions between meanings. In general terms, however, the research depends on the use of rewards and punishments to increase or decrease the appearance of a particular action. For example, if the experimenter's goal is to get a pigeon to spin in clockwise circles, he places the pigeon in a cage with a food chute that delivers seeds on the experimenter's command. As soon as the bird makes a slight clock-

The painting orangutan
Sydney, a two-and-a-half-year-old orangutan, is shown here with his teacher, Niall Ormund. Sydney's first painting was sold at Sotheby's on June 29, 1985, for 120 pounds sterling. This photograph was taken by his caretaker, Christopher Cormack.

The Minds of Animals: Behaviorism and Ethology

wise turn, the experimenter delivers some food. Once the pigeon has caught on and persists in making this slight turn to get its reward, the experimenter holds out until the bird turns farther, rewarding, say, a quarter turn. This procedure continues until the bird has to turn two or three complete circles to get its reward, and the patient experimenter has achieved his goal.

Behaviorism has made many important contributions to modern psychology. Its expansion to behavior therapy has proved helpful in some instances to disturbed children, mental patients, smokers who wish to quit, and persons who have an irrational fear of crowds, heights, or other particular situations. And behavior therapy can lead to faster, more direct results than alternatives like psychoanalysis. Despite the fact that behaviorism obviously works with laboratory animals and seems to be helpful in human situations, there are serious objections to some of its implications.

Critics become particularly harsh when it comes to Skinner's explanation of human language learning. He has asserted that children learn to emit the sounds of language in response to rewards in the form of parental attention and food much in the same way that the pigeon learns to spin in circles. As appears in the section on language, this view generally ignores a number of biological aspects of language. And in general, Skinner has been criticized for putting down the importance of biology, particularly the biology of the brain. In more recent writings, however, his antibiological stance seems to have shifted somewhat, allowing room in his scheme of things for research on the brain. His assertions that studies of the brain must be conducted with a view to the behavior of the whole organism are not to be taken lightly.

At the same time that American behaviorists were developing a science that largely ignored the idea that some forms of behavior might be inborn, a European school was emerging whose principal concern is, to the contrary, those patterns of behavior that are intrinsic to an animal. In contrast to the very artificial conditions in the laboratory of the behaviorist, the conditions of interest to members of this new school of ethology are those that occur in the animal's natural surroundings, the conditions in which the species evolved to survive. Here, they reasoned, they could study those aspects of an animal's behavior that are part of its genetic survival kit rather than such behaviors as pigeon spinning that have nothing to do with what pigeons do in the wild.

Two of the most famous of these ethologists were Niko Tinbergen and Konrad Lorenz, both of whom won Nobel prizes for their studies of the natural behavior of animals. Photographs of Lorenz swimming or walking, followed by young geese that seemingly think he is their mother have appeared in magazines throughout the world. The phenomenon known as imprinting is an important kind of learning that occurs in many species of birds and is a form of learning quite different from what the behaviorists study. Newborn birds have some genetic predispositions but will imprint on many differ-

The Minds of Animals: Behaviorism and Ethology

Elementary nervous systems
The hydra (1) is a very small marine animal with a tubular body and a mouth surrounded by tentacles. Its simple nervous system, a loose network of cells, contains no coordinating center where activity is organized. It is typical of the most primitive nervous systems. In the more highly organized jellyfish (2), in which food-capturing parts of the body are separate from the parts used for movement, there is some centralization of the nervous system. Nerve nets, used to control swimming, tentacle position, and feeding, flow together into rings, where considerable interaction occurs. Still further organization occurs in the nervous system of the earthworm (3). Clusters of nerve cells called ganglia are the organizing centers. From the brain in the worm's head, two parallel bundles of nerve fibers run along the animal's lower surface. Within each segment of the body, a ganglion occurs on each fiber bundle. Two small fiber bundles run between each pair of ganglia. The ganglia control the muscles and mediate the sense of touch in each segment.

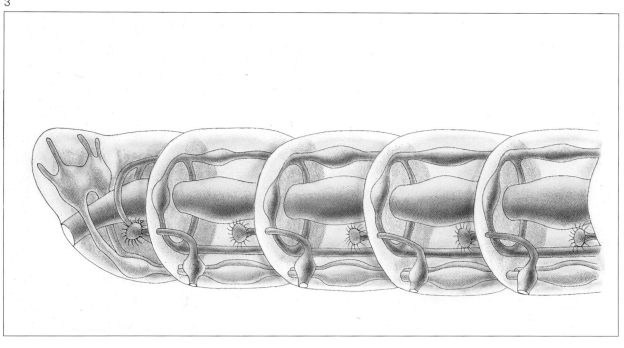

The Minds of Animals: Behaviorism and Ethology

1. Konrad Lorenz
Konrad Lorenz, along with Niko Tinbergen and Karl von Frisch, founded the field of ethology, the study of animal behavior in a natural setting. Lorenz was born in Vienna in 1903 and received the Nobel Prize in Physiology or Medicine in 1973. He demonstrated the phenomenon of imprinting, by which a newly hatched bird will become attached to the first moving object it sees. This is usually its mother, but Lorenz could imprint the birds on other moving objects, including himself.

2. A chimp and a baby lion
King, a four-year-old chimpanzee, has been taught to feed a two-month-old lion, Sandy.

ent sorts of things, whether their mother, a person, or a soccer ball, and they subsequently follow it around. So the youngsters have learned that the object in their environment has a particular significance, principally as a source of food.

Another important sort of innate behavior is called a fixed action pattern. In this case the animal goes through a stereotyped sequence of behaviors that occur whenever a well-defined set of circumstances occurs. Studies of fixed action patterns by American ethologist Daniel Lehrman and his successors have begun to show how they may be tied to brain function. Lehrman identified how hormones are involved in the unfolding of rather elaborate fixed action patterns in the reproductive behavior of ring doves. In such behavior the animal's past experience, sensory stimuli, and hormones are interwoven in controlling the brain mechanisms that govern behavior. Courtship behavior by the male dove causes the female's ovaries to grow and secrete hormones that trigger her mating behavior. The appearance of the eggs is accompanied by further hormonal changes that bring about incubation activities by both parents and the subsequent feeding of the young. Departure of the young birds leads to still other hormonal changes that cause the male to exhibit mating behavior, and the cycle begins again.

It is important that mutual interaction between partners coordinates the precise timing of events necessary for successful reproduction and that hormones play a key role in linking brain function and behavior. Lehrman's work, in fact, led to the demonstration that appropriate hormones injected directly into the brain can evoke specific components of the reproductive behavior. This powerful role of hormones in behavior has recently become important in efforts to preserve species of animals whose populations are declining. John Phillips and William Lasley of the San Diego Zoological Society have been using implants of the brain hormone known as gonadotropin-releasing hormone, a stimulant of reproductive functions, to try to provoke mating by green iguanas. Within three days of starting the hormone release in the females, the nearby males began to display behavior associated with mating. Whether this behavior is triggered by subtle behavioral changes in the females, a change in their odor, or some other stimulus is a puzzle yet to be solved.

(The complex links between hormones and behavior is explored further in chapter 12.)

Ethology and behaviorism begin to converge in the current work of Edmund Fantino, a student of Skinner's now working at the University of California, San Diego. He employs the analytical tools developed to study behavior in the laboratory to study naturally occurring behaviors. The behavior in question is foraging, the process of searching for food. Birds must expend energy in this process of searching for new sources of energy. Fantino's work concerns the birds' development of strategies to search for and select food under various kinds of natural circumstances. It applies rules for reinforcement of behavior developed in the laboratory.

Research in the behaviorist's laboratory often concerns a special type of activity known as operant behavior. This behavior involves a particular action on the part of the animal, like pressing a lever or a button. The action results in the animal receiving food or being affected in some other way that increases the likelihood that it will take the same action in the future. This procedure is known as operant conditioning, and the food or whatever presented to the animal as a result of its action is called the reinforcer. Experimental psychologists use a variety of operant-conditioning arrangements to study the effects of different kinds of reinforcement on such aspects of operant behavior as the frequency of the animal's action or the probability of its taking the action.

The results of studies of operant conditioning have had many successful applications to human problems in the area known as behavior therapy. Practitioners of this form of therapy attempt to identify actions that may be reinforcing undesirable forms of behavior or to encourage desirable behaviors by providing effective reinforcers. For example, compulsive collection of junk by inmates of a psychiatric ward was stopped when nurses stopped pestering the patients to remove the junk. The attention the patients received by collecting junk served as a reinforcer for the behavior. In the case of a woman whose undesirable behavior was to refuse all food, social contacts provided only when she ate served to reinforce the desired behavior.

In recent years much of Fantino's operant-behavior research has concerned the factors that influence making choices. We are

The Minds of Animals: Behaviorism and Ethology

almost always faced with choices to make from moment to moment in our lives. At this moment, for example, you can either continue to read this book, take a walk, fix a drink, or do any number of other things. What factors influence our decisions? Is it possible to use the highly developed technology of operant-behavior research to study how choices are made?

In an attempt to provide some answers, Fantino uses a procedure known as concurrent chains, developed to make sure that the only factor influencing a choice is how rewarding the outcome is. (The use of the term "rewarding" is not in keeping with the strict technical terminology of behaviorism, but for this example, it helps to think of a reinforcing presentation of food as somehow rewarding a certain behavior.) In Fantino's procedure, a pigeon has a choice of two white lights to peck. If it pecks the one to its right, a red light comes on after a variable interval of time. If the bird then pecks the red light, it is rewarded with some food, again after a variable time period. When the bird's first choice is the left white light, a green light comes on after an interval and the bird must then peck it to obtain food. With this arrangement it is possible to study how different delays in reinforcement after a colored light is pecked affect which white light is chosen more often.

Years of work with the concurrent-chain procedure has led Fantino to formulate what he calls a hypothesis of delay reduction to explain the behavior of pigeons in his choice situation. It proposes that the animal will prefer the white button that leads to the greatest reduction in time until it receives food. The hypothesis further predicts the relative frequency of the choice of one button or the other as a function of both the average time between food rewards and the time it takes to receive a reward after each button is pushed. The details need not concern us here. What is important is that the hypothesis predicts that the choice behavior will be affected in specific ways by specific changes in the timing of reinforcement.

For example, if the red light leads to food after an average of five seconds and the green light after an average of ten seconds, the red light is a slightly stronger reinforcer than the green light. If the white lights both lead to a turning on of a colored light after an average of five seconds, the bird won't bother at all to

1

2

The Minds of Animals: Behaviorism and Ethology

Animal models
Animals with artificially induced diseases that resemble human diseases are often used in pharmaceutical laboratories to test new drugs. Here a rat with a unilateral lesion in the brain region known as the substantia nigra provides an animal model of Parkinson's disease, a movement disorder arising from degeneration of the substantia nigra. The lesion causes the animal to turn in circles. The apparatus attached to the animal measures the effects of drugs on its turning behavior and thus indicates whether a drug is likely to aid patients with Parkinson's disease.

peck the white light that leads to the green light, even though there is a difference of only two and a half seconds in the average time to receiving food. The reason is that the choice of the light leading to the green light actually increases the average delay to food reinforcement by two and a half seconds, and Fantino's hypothesis says that delay reduction is the driving force. On the other hand, if the average delays after the colored lights turn on stay at five seconds for the red light and ten seconds for the green one, while the delay from the white lights to a colored light increases to two minutes, the bird will pick either white light with about equal preference.

What does choice behavior under these highly artificial laboratory conditions have to do with naturally occurring behavior that interests the ethologists? A great deal, according to Fantino, at least for certain sorts of behavior. One in particular that he has begun to analyze is the choices that animals make in foraging for food. Researchers who have studied foraging behavior propose that the law of natural selection (survival of the fittest) has led to the evolution of what is termed optimal foraging behavior. This means that if one kind of prey becomes scarce, an animal learns to seek other kinds of prey rather than spending time and energy seeking a prey that it may prefer to eat but can seldom find. In other words, it maximizes the rate of taking in energy in the form of food. Fantino proposes that his laboratory procedures of concurrent chains can be used to study the choice making involved in foraging and that his hypothesis of delay reduction is analogous to the notion of optimal foraging.

The proposed relationship between the concurrent-chain procedure and foraging relates the time a white light is on to the time an animal is searching for prey and the time the colored light is on to the time it takes to actually capture and consume the prey. Pursuing the examples given above, delay-reduction theory predicts that if the time taken to find food increases, an animal becomes less selective in what it chooses. It also makes other predictions such as that an increase in selectivity will be associated with increased capture time. Sometimes, however, optimal-foraging theory and delay reduction make different predictions about foraging behavior. Optimal foraging says that if there are two acceptable kinds of prey, one somewhat preferred over the other, the introduction of a third form of prey with an unacceptably long capture time would not affect the choice between the two acceptable kinds of prey. Delay-reduction theory says that even though the third kind of prey would not be accepted, its introduction would tend to reduce the preference of one acceptable form of prey over the other.

Even though there is much work to be done in evaluating the validity of operant-behavior research as a model for various forms of naturally occurring behavior, the preliminary success of delay-reduction theory is encouraging. We know that various forms of behavior have evolved because they have survival value. But we have little idea as to what factors actually influenced animal behavior in its evolution. If delay-reduction theory is correct, the behavior of foraging animals is influenced by factors that reduce the delay to reinforcement, not by the rate of energy intake or rate of reinforcement. Thus, behaviorism can actually contribute to our understanding of what guides behavior in the environment. It thus helps us to understand the interplay of the many factors that influence behavior: an individual's genetic heritage, experience, and the circumstances of the moment. The psychological laws of modern behaviorism can thus help close the loop between individual behavior, culture, and genetic change.

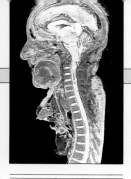

Cricket Love Behavior and the Cricket Brain

by Franz Huber, Max Planck Institute for Behavioral Psychology

During the course of phylogeny multicellular organisms evolved with different forms, metabolisms, and behavioral strategies, adjusted and optimized to the specific constraints of their habitat. Like us, these organisms are equipped with sensory inputs, central nervous systems, and output systems to execute behavior. Invertebrate animals, like earthworms, crayfish, insects, and mollusks, are often called simple organisms because they perform rather stereotyped behaviors with little or no learning. In addition, their nervous systems possess fewer nerve cells (several thousands to several hundred thousands and not millions as found in vertebrates and man). Also the number of connections among nerve cells is smaller. But the greatest advantage of such simple organisms for bridging the gap between behavior and nervous-system function is that one is able to identify single cells and to unravel their importance for a specific behavior. Furthermore, these invertebrate animals also teach us the general principles of nervous-system and brain function and thus help us to develop concepts and methods that may finally guide our approach to elucidating our own brain at the cellular and the system level.

Since ancient times crickets were known to communicate with sound signals produced by the male for finding mates. Females are attracted by these signals and approach the singing males, a response called "phonotaxis." Each species has its own love song, and the songs of different species differ in the sound-frequency spectra and the temporal organization of the song components (chirps and syllables).

This strategy is called "intraspecific acoustic communication" and involves the emission of species-specific signals by one member (the sender) and their reception and understanding by a second member (the receiver). The relationship between sender and receiver can be studied in two ways: by analyzing the love songs of the males and observing the phonotactic response of the females (the behavioral approach) and by studying the sensory cells, nerve cells, and muscles that take part in this communication (the neurobiological approach). As an example, I shall consider here the love song of the European field cricket *(Gryllus campestris)* and the response of the female.

As the sender, the male cricket produces its love song by raising the two front wings, each of which carries a file and a scraper. The file on one wing rubs against the scraper on the other wing, and sound is produced by the resulting fast oscillations of both wings. The rhythm of wing movements determines the temporal structure of the love song.

We know the way the wings are moved, the muscles driving the wings, the motor nerve cells driving the muscles, and higher-order nerve cells in the brain that command the song. We have learned that the love song is generated by a number of nerve cells within the thoracic ganglia (these are assemblies of nerve cells) and is controlled and commanded by nerve cells in the brain, which is the first ganglion in the head of the cricket. If one penetrates thin metal wires into distinct parts of the brain and stimulates these parts electrically, the male then starts to sing and keeps on singing. The song produced resembles the love song. The activation by electrical stimulation in the brain is carried by nerve cells that descend to the motor centers in the thorax, and there a set of nerve cells responsible for the rhythmic wing movements becomes active.

The female cricket is not able to sing; she does not have specialized front wings with file and scraper. But she listens to the love song and responds to it as soon as she is in a mating mood, which is controlled by hormones. Responding to sound requires hearing organs, ears, which in crickets are situated in the front legs. In the field a responsive female, upon hearing the love song, starts to walk and approaches the singing male. For this behavior, the female has to accomplish two equally important tasks: she must recognize the love song as belonging to her species and discriminate it from other sounds (the recognition task), and she must locate the sender (the orientation task).

A behavioral paradigm was developed in recent years, the walking compensator (a spherical treadmill) on which the female can freely walk and exhibit phonotaxis. With the female walking on the treadmill and by playing love songs to her, it was found that she extracts two messages out of the love song: the species-specific sound frequency band (4,000 to 5,000 hertz) and the species-specific rate of syllables (ranging from 18 to 40 hertz).

In our search for the sensory and nerve cells involved in this task, my colleagues and I first established the sense cells in the cricket ear that are most sensitive to the love-song pitch. The sense cells of the ears are not specialized to encode only the species-specific temporal pattern of the song. Thus, the ear cannot be considered an intimate part of the recognition setup.

The message of the two ears about song frequency, song direction and intensity, and song pattern reaches the central nervous system at the first thoracic ganglion. There we found several nerve cells, unique in form and function, that process this message. A pair of mirror image cells called omega cells (their shape resembles a capital omega) computes direction by processing input from both ears. Each member of the pair receives auditory input from only one ear and is inhibited by input from the other ear. This inhibitory interaction helps to sharpen binaural contrast and improves direction finding. The two omega cells can thus be considered to play a role in the neural basis of the orientation process.

However, females perform their phonotaxis only with an intact brain. So information from the nerve cells of the prothoracic ganglion must reach the brain. We found nerve cells whose axons ascend to the brain. These cells also receive input from the ears modulated by the activity of the omega cells, which indicates that the brain is informed about the current position of the female with respect to the sound source.

We have seen that cricket ears are not specialized to hear only the species-specific love song, so we looked for a recognition process in higher-order nerve cells. Such cells can only serve recognition if they respond to a frequency band of 4,000 to 5,000 hertz and, most important, if they specifically respond to syllable rates in the range of 18 to 40 hertz.

We were recently successful and discovered nerve cells in the cricket brain that fulfill these criteria. They receive song information via the ears and nerve cells in the prothoracic ganglion. Among brain cells, we found some that preferentially react to syllable rates between 18 and 40 hertz and that thus mimic the range of the phonotactic response of the female when walking in the field or on the treadmill.

Cricket love behavior, expressed by the love song of the male and by the orienting response of the female, requires the brain as an important part of the neural system controlling song production and song recognition.

1. Male cricket
A male cricket raises his anterior wings and rubs them together to produce its mating call.

2. Female cricket
In this experimental arrangement the female walks on a rotating sphere driven by motors. Infrared light reflected from the white circle on her back controls the motors so that the sphere precisely compensates for her movements, keeping her in the same position while she walks. This arrangement permits accurate measurement of her response to various mating calls.

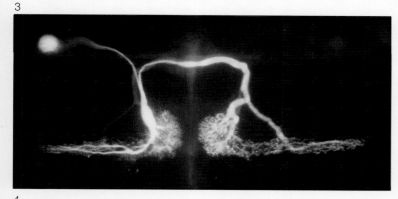

3-4. Omega cells
Omega cells have been injected with a fluorescent dye that fills all their fine branches. The body of the omega cell (the circle in the upper left corner of figure 3) sends a fine nerve fiber to the structure shaped like a capital omega in the center of the figure. This cell is activated by sound coming from the left ear, which is located in the anterior leg on the same side as the cell body. Nerve signals from the ear are received by the branching fibers at the lower left and cross to the opposite side of the body, where they inhibit the activity of the omega cell on that side. (The midline is indicated by the fuzzy vertical line.) This arrangement, found in both males and females, permits the cricket to determine the direction of a sound source. Omega cells occur in mirror image pairs, one on either side of the body (4). Each cell receives nerve signals from the ear on its side of the body and inhibits the cell on the opposite side.

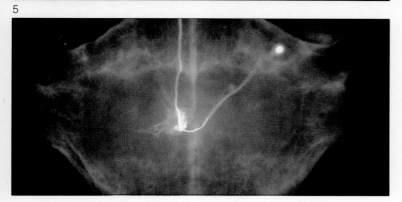

5. The prothoracic ganglion of the female
This view of the prothoracic ganglion shows the body of a nerve cell in the upper right. It sends a nerve fiber down toward the center and across the midline, where it produces a small field of fine branches. It then continues to the front part of the brain. This neuron is particularly sensitive to sound at the frequency of the mating call (4,000 to 5,000 hertz), but it does not distinguish mating calls from other sounds.

6. The cricket's brain
This diagram represents the left half of the cricket's brain, bounded by a heavy black line. An ascending nerve fiber from the prothoracic ganglion (AN-1) brings nerve signals originating in the ear. The branches of this fiber connect to the branches of the green brain cell (BNC-1) toward the upper surface of the brain. Branches from the green cell in turn connect with those of the red cell (BNC-2) in a more central region. Thus signals arriving from fiber AN-1 activate BNC-1, which in turn activates BNC-2. The bottom line of the lower left diagram represents four syllables (bursts of sound) of a mating call. The three lines above are the electrical signals recorded from each of the nerve cells as the syllables occur. Cell AN-1 responds to each syllable after a short delay (the first two spikes are spontaneous activity). Cell BNC-1 responds with a longer delay and without a clear relationship to the syllables. The relationship to the syllables is completely lost in the still later response of cell BNC-2. While all the cells respond to sound at frequencies corresponding to those of the mating call, only cell BNC-2 distinguishes the mating call from other sounds.

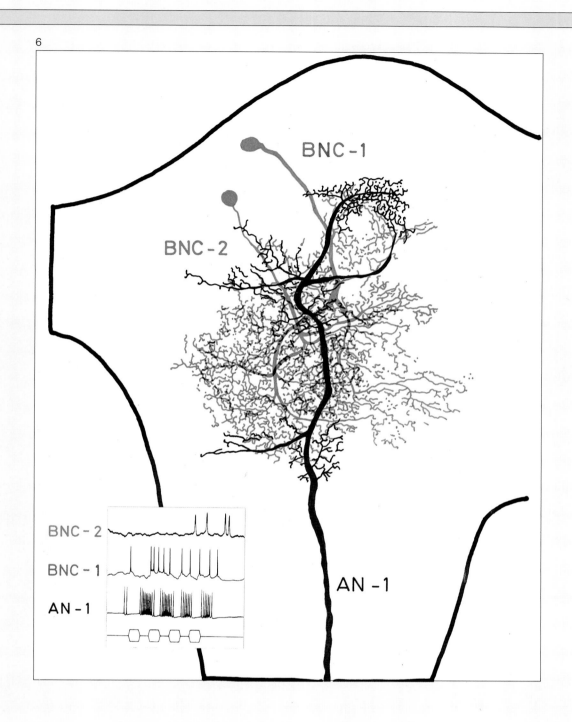

4
Gender, Culture, and Genes Make a Difference

There are many obvious biological differences between men and women, virtually all of them having to do with reproductive functions. For years people have debated whether or not there are intrinsic differences in behavior. Do little boys play with toy trucks while little girls play with dolls because of some basic biological difference, or is it a consequence of cultural effects? Do girls have a naturally better verbal ability than boys, and do boys tend to excel in spatial skills? Despite all the attention these questions have received during recent years, there is still far more heat than light to the debate. If there is any truth to such assertions, there might be some difference between their brains. Males and females do have different balances of hormones flowing through their bloodstreams, and hormones do have powerful influences on many organs in the body. Doreen Kimura of the University of Western Ontario has recently shown that women's skills do change during the menstrual cycle. When estrogen is low, spatial skills are at their best, while complex motor tasks are at their worst. High estrogen levels reverse the picture.

Some definite differences have been found between the brains of male and female animals. In the songbird, for example, Fernando Nottebohm at the Rockefeller University finds that the parts of the male brain responsible for producing songs are significantly larger than the corresponding regions in the female brain. The search for differences between male and female human brains over the course of the past century has often become snarled in arguments over male versus female superiority. On the average, male brains are larger than female brains, but there is no known reason why a slightly greater brain mass should confer greater mental powers upon its owner.

Some tests indicate that the two halves of the female brain function more symmetrically than those of men, which suggests that there is less specialization of function in the hemispheres of the female brain. In most individuals, the left hemisphere is specialized to deal with language, while the right dominates for spatial abilities. The notion of a difference in symmetry gets some support from the fact that the part of the left hemisphere of the male brain that is responsible for some language functions is larger than the corresponding region of the female brain. In addition, the bundle of nerve fibers that carries information between the two halves of the brain, the corpus callosum, is larger in females than in males.

A quite recent report of differences between male and female brains come from D. F. Swaab and E. Fliers at the Netherlands Institute for Brain Research. They studied a region of the brain known as the hypothalamus, which is an important control center for maintaining such fundamental life processes as reproductive behavior, eating, drinking, and regulation of body temperature. A group of cells in this area of the brain of laboratory rats had been found to differ in size between males and females. Swaab and Fliers examined a corresponding region in the human brain and concluded that it is about 2.5 times larger in males than in females. What this means in terms of brain function is not clear. The cells are located in an area that seems to be related to sexual functions, but just what they do we still do not know.

The finding that male and female brains exhibit differences, albeit rather minor ones, raises the possibility that there might be cultural differences between brains. After all, as this book repeatedly points out, experience, particularly experience early in life, can affect the way the brain is organized. Do differences in culture, particularly dramatic differences such as those that exist between East and West, show up somehow as differences between the brains of the oriental and the occidental? As we have seen, finding differences in human brains that relate even to such dramatic differences in capabilities as intelli-

Hormones
Athletes who use male sex hormones to enhance their performance are a growing problem in the world of sports.

gence is a task fraught with difficulties. The same is true with cultural differences, but some scholars insist that they must exist.

If brain differences exist between peoples of different cultures, how might culture and biology influence one another? In a sense, this question is another way of asking about the biological nature of mind, since culture is a product of the collective mind of a society. Over the years this question, which really concerns the interactive evolution of both mind and culture, has troubled biologists, anthropologists, and social scientists. We are still very far from having an answer, just as we are far from having key information about the biological basis of the individual mind. Once again, however, the clue to progress may lie in learning how to ask the right questions.

One prominent source of new questions about the relationship among minds, cultures, and biology is a rather new branch of science known as sociobiology, founded by Edward O. Wilson, a biologist at Harvard University. In general terms, Wilson was trying to develop a theory that could explain the biological basis of social behavior in animals and in humans. A particularly challenging aspect of social behavior is an explanation for altruistic acts. If an individual acts to save himself from an enemy, the obvious motivation is self-survival. If a mother bird endangers herself to save her young from attack by a cat, Wilson explains her behavior by noting that it enhances the likelihood of the survival of her genes carried by her young. If the mother were to save herself and allow one or more of her young to be killed, her genes might disappear from her species gene pool, a process that would also eliminate the genes responsible for her selfish behavior.

Taking this argument one step further, Wilson proposes that altruistic behavior, such as one bird making a warning cry when it sees a circling hawk, similarly serves to enhance the chances that an individual's genes will survive. The warning cry increases the likelihood that the crier will be attacked, but it also makes it more likely that the flock as a whole will survive. And since the flock contains brothers, sisters, and other relatives of the crier, at least some of his genes will survive in these members, even if he should perish. Wilson's theory of gene survival pertains to humans as well, in his view, and, he claims, it explains loyalty to a group and acts of heroic self-sacrifice in the name of this loyalty.

In a more recent extension of the ideas of sociobiology, Wilson was joined by Charles J. Lumsden of the University of Toronto in an attempt to explain the interaction among genes, mind, and culture. They propose a closed loop of influences that tie together the evolution of these three attributes of human beings. Genes, they propose, direct the formation of the human brain, giving it a propensity to think in certain ways and not in others. The examples they cite include accepting certain ritual or sexual practices while rejecting others. The resulting cultural behaviors in turn influence the evolution of genes, favoring the survival of genes that play a role in the generation of certain behaviors, while leading to the elimination of others. This process is continuous, of course, with genes constantly changing as new brain patterns evolve that lead to new behavior patterns and so on in an endless cycle.

There is a certain amount of appeal in this idea, but it is also a very controversial one. It would be difficult to support a different theory that says that genes evolve independently of culture, or that culture alone evolves independently of biological influences. The Wilson and Lumsden theory does, however, make certain assumptions about the identity of, or at least close correlation between, mind and brain, which we have already seen to be controversial. But more than this, sociobiology suggests that there are differences in genetic fitness between races or nations and that human behavior is determined, implications that many critics find unpalatable. The proposed links between genes, minds, and cultures are provocative. They will surely continue to stimulate scholarly debate for years to come. The Wilson and Lumsden theory will require testing by much careful scientific research before we can be sure of what it really tells us about how the human mind and brain relate to our genetic makeup and the great variety of cultures man has created.

Gender, Culture, and Genes Make a Difference

1. Egrets displaying before mating
The male displays his feathers, while the female bends her neck as a sign of submission.

2-3. Symbols
A traditional view of the difference between the sexes is emphasized in these drawings made by children with minor cerebral dysfunction, studied by Katerina Michelsson at the University of Helsinki. Small boys play with guns and toy cars, while little girls dress up. Does this attitude reflect some difference in brain structure between men and women, or is it a cultural artifact?

Gender, Culture, and Genes Make a Difference

1. The two halves
An engraving by the German artist Uwe Brandi entitled Zwei Halbe, or the two halves. It shows the machinery of the two hemispheres of the brain in cartoon form.

2-3. Nobel babies
Can the genius of Nobel laureates or other gifted individuals be transmitted by artificial insemination? This is a question posed by the birth of the Californian Nobel baby Doron Blake (2). Will he follow in the footsteps of these famous Nobel prize winners gathered together in Brussels at a 1927 scientific conference (3)? Scientists remain skeptical.

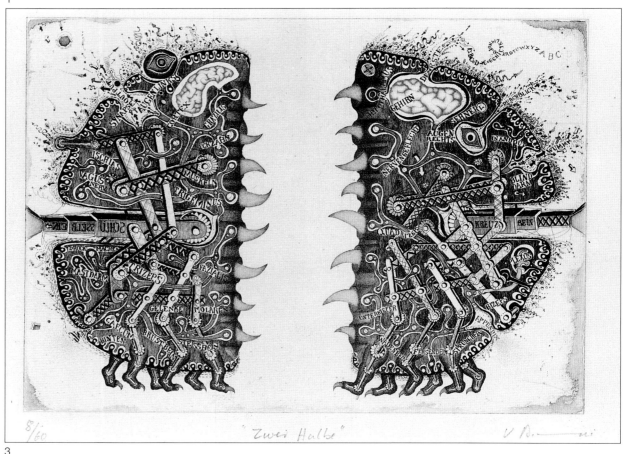

Gender, Culture, and Genes Make a Difference

1-2. The samurai and the geisha
Do differences in civilization reflect differences in the brain? Is it possible to speak, for example, of an American brain and a Japanese brain?

3. Young geniuses
This school in Philadelphia, founded by Glenn Doman, has as its motto "Every child can become a genius."

Gender, Culture, and Genes Make a Difference

Yin and yang
A Chinese symbol of the complementary dualism of the universe: feminine and masculine, dark and light, earth and sky, etc. The black region is yin, the red yang, contraries that need not be in opposition.

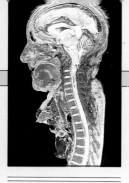

The Brain and the Medical Traditions of the Eastern World

by Jean-Claude de Tymowski, M.D., president of the International Medical Alliance and director of the National School of Acupuncture

Many plants and animals have no nervous system. However, this does not prevent them from being highly structured and possessing a highly developed capacity to adapt to the environment.

Eastern theories of medical physiology, in particular Chinese Taoist medicine, hold that man is a spatial and temporal microcosm of the universe, that our cells retain the memory of their origins at the beginning of the universe, and that our cells and certain clusters of cells in the human body retain this property of self-structuring, which is independent of the nervous system.

Tibetan philosophy quite clearly expresses the idea that the ontogenesis of the human body is based on the ontogenesis of the world and is engraved on the different organs of our bodies and also in human dental structure. According to this theory, the first stage of creation, that of cosmic forces, corresponds to the bones of the shoulders and the pelvis, the first budding of the limbs, and the first incisors. The second stage of creation, that of masses, corresponds to the arms, the thighs, and the second incisors. The third stage, the creation of plants, corresponds to the forearms, the lower leg, and the canines. The fourth stage, gender, corresponds to the carpus, the tarsus, and the first bicuspids. The fifth stage, the creation of animals, corresponds to the metacarpus, the metatarsus, and the second bicuspids. The sixth stage, the creation of man, corresponds to the first phalange and the first molar. The seventh stage, the creation of mind, corresponds to the second phalange and the second molar. The eighth stage of creation is known as "the creation of the three principles": the trinity, the divine, and the sublime. It corresponds to the phalangette or ungual phalanx, the phalange associated with dexterity, and the expression of the sublime in art. It is interesting to note that the eighth molar, the last molar to appear, is referred to in both Eastern and Western traditions as the wisdom tooth, which shows that certain aspects of Eastern philosophy either penetrated into Western traditions or originated spontaneously in both cultures.

This theory of cosmogony traces the evolution of life and of the world. Animals belong to the fifth stage, which corresponds to the appearance of the central nervous system and the functions of the brain. For Eastern philosophers, life was able to evolve up to this point without a brain, and it is in fact this concept that constitutes the major difference between Western and Eastern traditions.

In Western medicine, the brain and the central nervous system symbolize the coordination of life and play an essentially centralizing role. The brain is also considered to be the seat of individual personality, so much so that Western medicine considers that life has ceased when brain activity ceases and that a flat electroencephalogram corresponds to clinical death, even if all other vital functions can be maintained.

In general, Eastern medical traditions have a radically different concept of this question. They assign a particularly important role to the other organs and consider that they play a far greater role than is admitted by Western medicine. Each individual organ or set of organs is considered to be the seat of one of the so-called plant souls and possesses emotional properties, or properties that govern character traits or behavioral patterns, which Western physiology considers to be controlled by the brain. In the Eastern medical tradi-

51

tion, behavioral patterns are related to the five major organs, and the fact that a part of the organization of life took place long before the appearance of the central nervous system is retained in the body's memory.

Eastern philosophers believe that life is the result of an active fertilizing principle that generates movement, referred to as yang, and a reactive materializing principle, referred to as yin. The human body is a matrix that receives the incarnation of entities: the *shen*, which exist prior to birth. The heart is the first abode of the *shen*; in other words, the vital principle resides in the heart. It follows that some of the feelings or emotions that Western medical traditions associate with the brain are in Eastern medicine associated with the heart, the seat of the soul. Recent Western embryology has shown that the heart begins to beat very early in embryonic life. This idea is echoed in popular or esoteric Western traditions, which tend to associate many feelings or emotions with the heart.

In Eastern thought, such activities as reflection, pondering, and sympathy, which carried to excess may predispose the individual to anxiety, are psychic predispositions linked to the spleen and stomach and to the solar plexus. This explains how psychic predispositions can result in somatic disturbances in the corresponding parts of the body, as in the case of anxiety, which can cause stomach ulcers. Melancholy, sadness, and a certain kind of neurotic, depressive state are believed to be related to disorders of the lungs and large intestine.

The kidneys and the bladder play an important role in Taoist medicine. The kidneys are considered to be the seat of ancestral energy, which is responsible for reproduction, a concept echoed in the Biblical phrase "the issue of his loins" to describe a man's descendants. In Taoist physiology, the same word is used to refer to the substance within the spinal cord, bone marrow, brain tissue, and sperm, all of which share the characteristic of softness.

Tantric practices are designed to reinforce the energy of the so-called sea-of-marrow by affecting the relationship between sexuality and the brain, as in the practice of kundalini yoga, in which tantric exercises are believed to increase and unify the energies of the upper and lower body, particularly the sea-of-marrow region. The brain is believed to be nourished by currents and by the passage of energy.

The Chinese ideogram nao, which stands for the brain, is made up of three parts, one of which is the symbol for running water and also for the hair. Chinese medicine may thus have been aware of the existence of brain waves, which modern Western science detects by means of electroencephalograms. The second part of the ideogram represents a box, the skull, containing a cross dividing it into two. This may well represent the different hemispheres of the brain. The third part of the ideogram has given rise to some controversy: some experts believe that it symbolizes the soft brain tissue, while others believe that it relates to the ideogram for man.

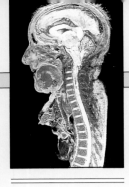

Hemispheric Dominance in Japan and the West

by Tadanobu Tsunoda, Auditory Disorders Department, Medical Research Institute, Medical and Dental University of Tokyo

As a part of my research on aphasia and speech disorders due to cortical lesions, I have conducted over the past fourteen years comparative studies of cerebral dominances in Japanese and non-Japanese subjects. I should emphasize here that these studies involved people with perfectly healthy brains. Many earlier lateralization studies have involved bisected brains in which the right-left connections had been surgically cut. In such studies one cannot observe an essential region in which switching occurs, which is a key factor in my research.

Here I shall discuss the predominance of the left hemisphere for speech, calculation, and logic and that of the right hemisphere for music, recognition of shapes, and all that pertains to analogy and the emotions. I have developed a new method for studying dominance based on delayed auditory feedback, close to the way a speaking man might hear his own voice. I have observed some surprising differences between Japanese and occidentals.

In order to understand what follows, one must be aware of one of the peculiarities of the Japanese language. A complex sentence with a well-supplied vocabulary can be expressed using only vowels. The sentence "Ue o ui, oi o ōi, ai o ou aiueo" has a meaning and can be translated as "Worrying about hunger, disguising his old age, he chases love, a love-starved man." Individuals whose native language is Japanese, therefore, have a very special habit of employing vowels.

What can we observe from all this? The left hemisphere, of course, receives speech sounds, including emotional voices, but it also receives the sounds with a similar structure, like the noise of waves, wind, and rain, the babbling of a brook, and all the noises of nature. The left hemisphere picks up nonharmonic sounds, while the right hemisphere dominates for harmonic sounds.

One of the most striking examples is that of music. The right hemisphere dominates when Japanese listen to the sounds of such occidental musical instruments as the piano, the violin, or the contrabass, and the left hemisphere dominates for the sounds emitted by a *shō*, a *shichiriki*, a *biwa*, and a *shakuhachi*, which are traditional Japanese musical instruments. This phenomenon has been recently verified by Y. Kikuchi and T. Tsunoda, who looked at auditory evoked potentials. In these experiments the subjects could not recognize the instruments used because the sounds were so brief (between 1/40th and 1/13th of a second).

The automatic switching between the two hemispheres is ensured mainly by the brain stem, located under the cortex; it divides the brain into the harmonic brain (right) and the nonharmonic brain (left). Among other things, the tests revealed that the switching function is determined up to the age of nine by the exposure to speech sounds in a certain language. For most non-Japanese (except the Polynesians) the sounds perceived by the left hemisphere (the side for speech) are actually limited to syllables containing consonants, while vowels, the noises of our environment, and the sounds that express an emotion ("Oh!" "Ah!") are perceived by the right hemisphere. More generally, emotion is perceived by the right hemisphere for non-Japanese, who use many more consonants in their speech, and by the left hemisphere for Japanese, who employ many vowels. In the same way, cerebral dominance for sexual functioning, which is extremely laden with emotion, seems to be located in the left hemisphere for Japanese and in the right hemisphere for occidentals.

It was proved that these differences in brain patterns are caused by the environment and not by any racial difference. The structure of the brain, determined by the exposure to the language, remains unchanged after the age of nine. It is linked with the cultural environment and the development of sentiments as expressed in the spoken language. Thus, a Japanese child raised as an American until the age of nine will henceforth perceive his emotions like a non-Japanese, predominantly in the right hemisphere, after his return to Japan.

The Tsunoda key tapping method enables one to test hemispheric dominance for pure tones ranging from 20 hertz to 16 kilohertz and has led to an investigation of the 20-to-200-hertz range in great detail. The human auditory system may be divided into two parts: one part processes sounds of 100 hertz or higher, the frequency range to which all verbal sounds belong, and the other part processes sounds below 100 hertz. The detailed tests for pure tones from 20 to 200 hertz indicated that the auditory system works differently for pure tones of 100 hertz or more and for those below 100 hertz. For the former, higher-frequency group of pure tones a left ear advantage was observed, but for the latter, low-frequency group a right ear advantage was observed. Since sounds of 99 hertz or less are too low in frequency to be verbal sounds, the peculiar response of the auditory system to these low-frequency sounds may signify a subcortical biological function.

Dominance changes were observed for some still unknown reason at frequencies that are exact multiples of 40 and 60, e.g., 40, 60, 80, 120, 160 hertz, etc. Another dominance change might occur for a sound with a frequency equal to the exact age of the subject who hears it. The result of an investigation conducted on more than 30 healthy subjects has revealed that most of them exhibited this change on their birthdays and some subjects exhibited it over three consecutive years. It seems that this change in dominance takes place at a constant rate of revolution.

The switching mechanism is also closely related to the earth's revolution, lunar motion, and possibly other cosmic activities. The switching mechanism shows irregular shifts whose causes are still unaccounted for. The link with cosmic activity suggests that there is a miniature cosmos in the human brain. Yet we have lost our ability to perceive this microcosm within ourselves in the hustle and bustle of civilization. Prehistoric man, awed by the working of nature and also having a great deal of insight into nature's activities, was probably able to feel this internal cosmos.

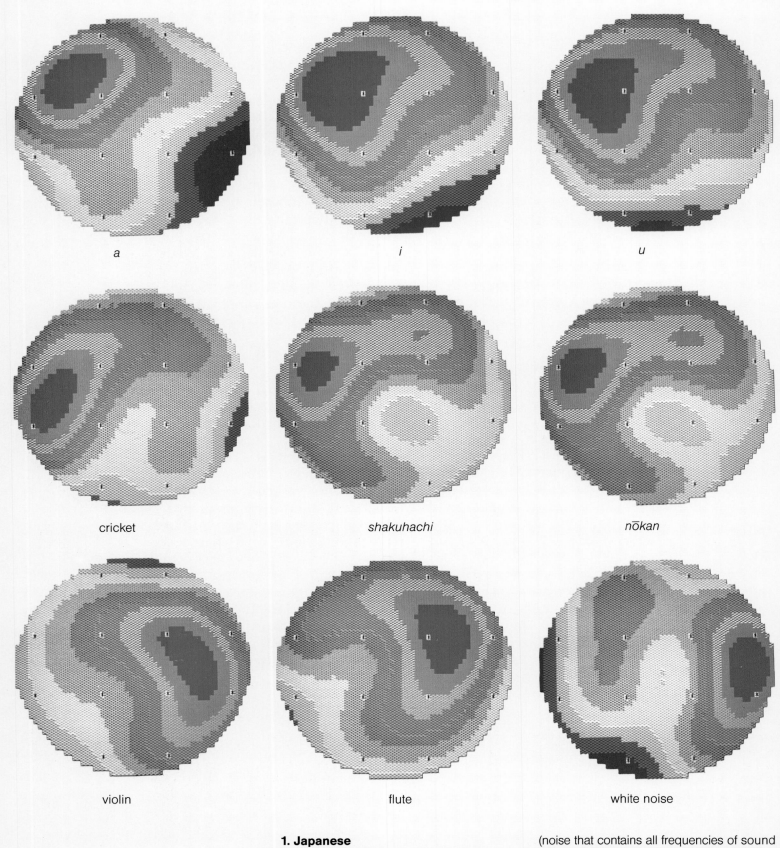

1. Japanese
The left hemisphere dominates when listening to the vowels *a, i,* and *u;* to the chirping of a cricket; to the sound of the *shakuhachi* (bamboo flute); and to the *nōkan* (another Japanese instrument). The right hemisphere dominates when listening to the violin, the flute, or white noise (noise that contains all frequencies of sound and resembles the noise made by a television set tuned to an empty channel).

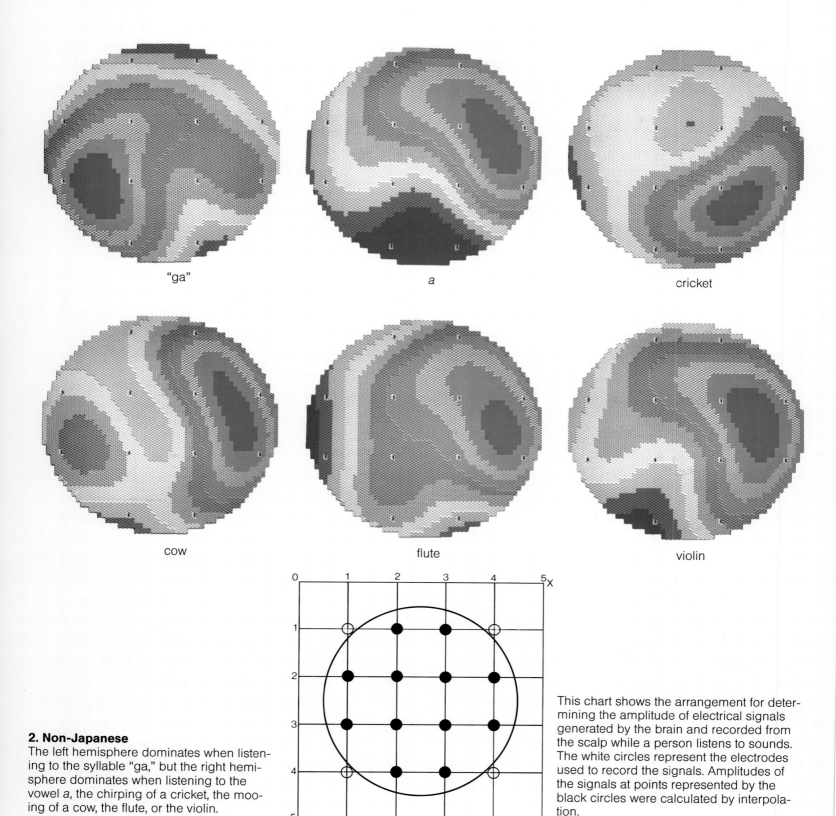

"ga"   a   cricket

cow   flute   violin

2. Non-Japanese
The left hemisphere dominates when listening to the syllable "ga," but the right hemisphere dominates when listening to the vowel a, the chirping of a cricket, the mooing of a cow, the flute, or the violin.

This chart shows the arrangement for determining the amplitude of electrical signals generated by the brain and recorded from the scalp while a person listens to sounds. The white circles represent the electrodes used to record the signals. Amplitudes of the signals at points represented by the black circles were calculated by interpolation.

57

Differences in cerebral dominance

Japanese model

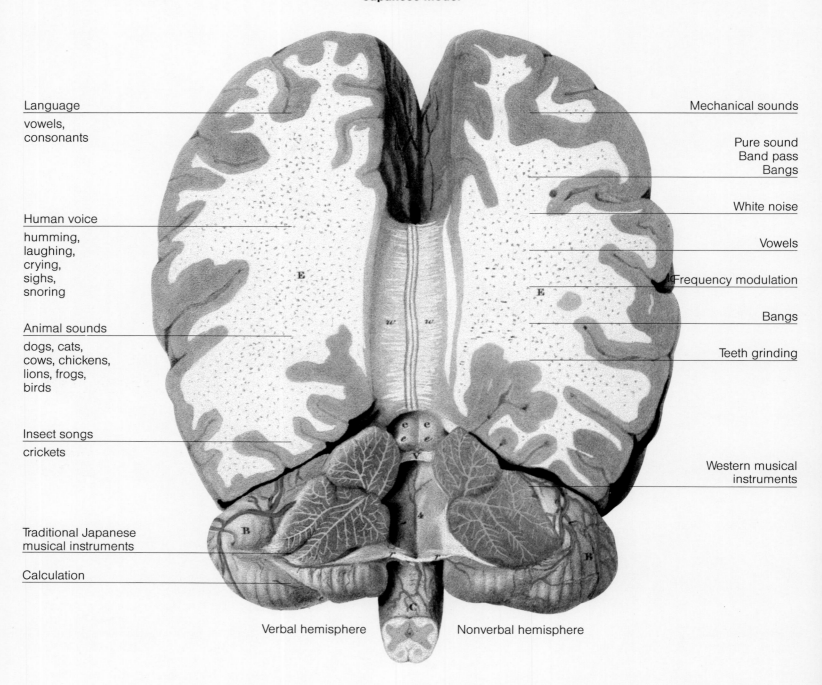

Language
vowels,
consonants

Human voice
humming,
laughing,
crying,
sighs,
snoring

Animal sounds
dogs, cats,
cows, chickens,
lions, frogs,
birds

Insect songs
crickets

Traditional Japanese
musical instruments

Calculation

Mechanical sounds

Pure sound
Band pass
Bangs

White noise

Vowels

Frequency modulation

Bangs

Teeth grinding

Western musical
instruments

Verbal hemisphere   Nonverbal hemisphere

between Japanese and occidentals

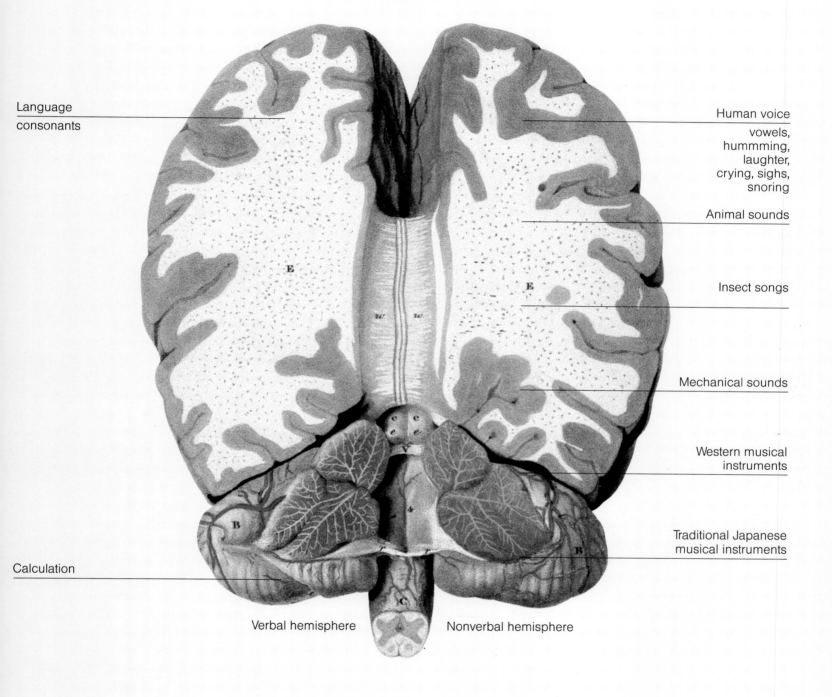

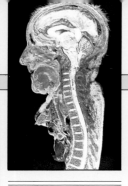

Sex Differences in Human Brain Organization

by Doreen Kimura, Department of Psychology, University of Western Ontario

It is to be expected that men and women, so different in their adult behavior, might have different brain organizations. Yet we are at the very beginning in our understanding of what these differences are. Even research with nonhuman animals has only recently begun to uncover the brain mechanisms in the hypothalamus that underlie male-female differences in such basic functions as mating behaviors. Two lessons emerge from the animal literature. One is that the development not only of genital sexual differentiation but also of the brain mechanisms that control sex-appropriate behaviors is determined by sex hormones early in intrauterine life. The second is, rather surprisingly that basic gonadal structure and hypothalamic regulation of sex-specific behavior are asymmetric.

Such facts emerging from the animal literature provide indirect support for the notion that the human brain is sexually differentiated. It appears that brain organization in male and female humans differs very early in development and is determined in part by fetal hormones. Such differentiation is by no means limited to hypothalamic structures. In fact, studies of the neuroanatomy of human fetal brains suggest that sex differences are pervasive throughout the cerebrum, though at present we do not understand their full significance. However, there is also good evidence that sex hormones constitute one significant factor in the superiority that men on average show in some aspects of spatial ability. It seems probable that hormonal factors act in part by altering brain organization.

Jerry Levy proposed several years ago that females preferentially develop the left side of their bodies (including the left cerebral hemisphere), while males either showed no such preference or developed the right side (including the right brain hemisphere). This hypothesis was meant to account for the earlier development of speech and articulatory skills in girls (skills presumed to depend critically on the left hemisphere), while boys showed more development of spatial skills, dependent on the right hemisphere.

Later theories about sex differences in cerebral lateralization suggested instead that males' brains were generally more asymmetrically organized than females' brains. Thus, it was proposed that both speech and spatial functions were more bilaterally organized in females, while in males they were more dependent on the left and right hemispheres respectively than was the case in females. This scheme was proposed to explain the occasional finding in perceptual asymmetry studies that males showed more asymmetry than did females, which presumably reflects more asymmetric cerebral organization.

However, when one examines the effects of damage to one hemisphere of the brain in human adults, it is clear that gross speech disorders (aphasias) are not more frequent after right-hemisphere damage in women than in men. This argues against the idea that basic speech functions are more bilaterally or diffusely organized in women. In fact, there is recent evidence that basic speech functions depend on more delimited regions in the left hemisphere of women than of men, and thus, that such functions are more focally organized in women, not more diffusely.

Other verbal functions that go beyond basic speaking ability (such as the ability to generate words beginning with a particular sound or letter) appear to be organized very similarly in males and females. Still other, more abstract verbal abilities, such as vocabulary (word definition), do indeed show a more bilateral pattern of representation in women than in men. The picture that is emerging from ongoing research is that not all verbal functions are organized in the same way. Depending on which intellectual function one is examining, women's brains may be more focally, equally, or more diffusely organized than men's brains. Less is known about brain organization for spatial ability than for verbal ability, but one might expect a similar principle to operate.

5
It Takes Timing: The Rhythms of Life

What is the relationship among human sleep patterns, the fact that a person's body temperature reaches its lowest point about four in the morning, and birds flying south for the winter? The answer is that these and many other biological phenomena all involve fairly precise timing. This timing depends on something usually called a biological clock, a timekeeping system that depends on some brain activity, a brain activity that, despite many years of research, has still not been precisely pinned down. Anyone who has ever experienced jet lag after a long flight knows how important biological clocks are and what a strong influence they can exert on our lives. Despite a shelf full of books on how to overcome jet lag, however, we still do not know enough about biological clocks to be able to reset our own regularly and reliably.

There are a variety of cyclical events that occur in the body on different time scales. Most noticeable is the daily cycle of sleeping and waking, the so-called circadian cycle (from *circa*, "about," and *dies*, "day"). Some cycles are shorter. For example, the brain emits hormones that control the release of sex hormones and growth hormone in pulses that occur at intervals of about one to three hours. Other cycles are longer than the circadian cycle. The menstrual cycle repeats itself every 28 days. While there is little clear-cut evidence for annual cycles in humans, many lower animals do follow seasonal cycles of reproductive activity.

One phenomenon recently discovered by Norman E. Rosenthal, David A. Sack, and their colleagues at the National Institute of Mental Health in Bethesda, Maryland, offers some evidence for seasonal variation in human behavior. These workers found a group of people who become depressed in the fall and winter, becoming fatigued and lethargic, oversleeping, and overeating. In the spring and summer they return to normal or even become manic. Reasoning that day length is the most reliable and consistent seasonal cue and the most important cue for the seasonal cycles found in a variety of animals, they began to experiment with exposure of their patients to light. They found that extended exposure to light several times as bright as that required to read does indeed significantly relieve the patients' depression. Precisely how light exerts such antidepressant effects is not known, but, as will be discussed below, at least one brain chemical is known to be affected by exposure to light. The potential therapeutic applications of bright light are sure to stimulate much more research in this area.

It is not altogether certain whether the various cyclical activities in the body are under the control of a single biological clock or several different ones. Most experiments suggest that when one function is artificially shifted by a transoceanic flight or some other means, all other functions shift accordingly, which in turn suggests that just one clock governs all systems. On the other hand, experiments by Jurgen Aschoff at the University of Munich have shown that in some individuals there can be a dissociation between the cycles of body temperature (which normally reach a low during sleep) and the sleep cycle itself. One explanation is that there is not just one clock. Another is that more complex sleep-regulating factors are under the control of a single clock. The definitive experiments are yet to be done.

We are not particularly aware of the running of our clocks, but we are sometimes aware of their effects. Generally, a sleepy feeling comes upon us around the particular time of day when we normally go to bed. Then we tend to wake up at about the time the alarm clock is to go off, even on days when we have not set the alarm. So even though we are not conscious of our clocks, they do have a close relationship to what goes on in our minds. The power this timer in our brains can exert in our lives is most evident today to the traveler suffering from jet lag, but it was known long

The space mechanic
In space, the normal cues that keep the body's clock on time are missing, so cycles of sleeping and waking are disrupted. This may interfere with the ability to perform demanding tasks.

It Takes Timing: The Rhythms of Life

Sleep and dreams
Works of art often depict different states of the human body, like poses of relaxation, reflection, vigilance, and sleep. Fabienne Laffont who works in Henri-Pierre Cathala's laboratory of sleep disturbances at the Salpêtrière hospital has identified some of these states in the works shown here.
1. **Night,** *detail of Giuliano de Medici's tomb by Michelangelo at the church of San Lorenzo in Florence. It shows a woman in a state of relaxation and deep thought.*
2. **Sleep of the Apostles,** *mosaic of the Dome of Monreale in Sicily (twelfth century). Depicted here is a state of deep sleep.*
3. **Nightmare Leaving Two Sleeping Women,** *by J. H. Fussli, Kunsthaus, Zurich (1870). This illustrates the typical disturbed awakening that follows a nightmare.*
4. **Dream Caused by the Flight of a Bee around an Apple One Second before Awakening,** *Salvador Dali, Lugano-Castagnola (1944).*
5. **Old Spinner Asleep,** *by G. F. Cipper, called Todeschini (thirteenth century). The painting shows the lapses of alertness that affect the aging brain.*

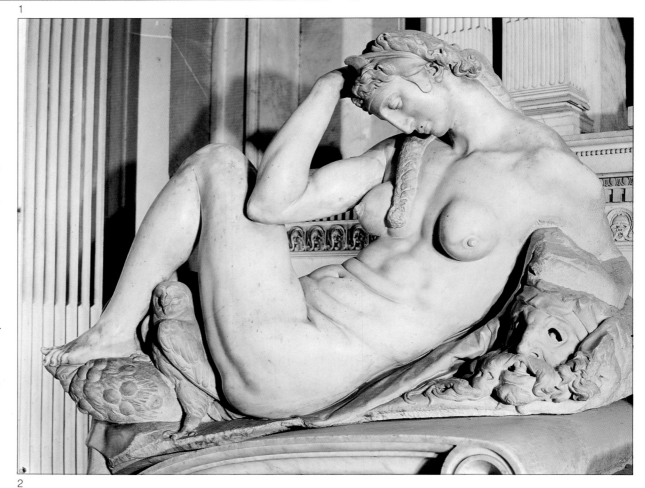

It Takes Timing: The Rhythms of Life

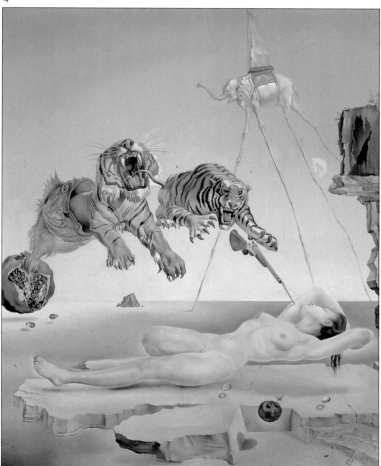
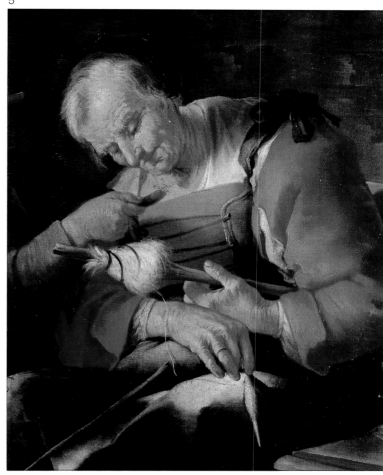

before by factory workers who were transferred from the day shift to the night shift or vice versa. Today we are beginning to learn a bit about the potentially damaging effects on health that may result from the abuse of our body clocks by such practices. Researchers are also becoming aware that shifts in the setting of the clock occur during aging and can be related to some of the unpleasant experiences of aging, like the disruption of sleep patterns.

If people are placed in isolation chambers without any real physical clocks and no clues like the position of the sun to help them know what time it is in the outside world, their free-running body clocks tend to generate 25-hour cycles. It seems that exposure to the cues associated with the 24-hour cycles of day and night is necessary to keep the body clock synchronized. Exposure to light does seem to have something to do with keeping our clocks on the proper time. But how does light affect the brain?

One of the principal components of the time-keeping system of the brain is a small organ known as the pineal gland, which lies buried beneath the brain's two hemispheres. The French philosopher René Descartes considered the pineal to be the seat of the soul, largely because it is the one and only part of the brain that does not appear in duplicate on the two sides of the brain. While the notion that the pineal gland contains the soul is not terribly popular today, we still do not know very much about just what it does do. Even some very thick, advanced physiology textbooks do not bother to mention it. One thing it does do that seems to be increasingly important in understanding our body clocks is to produce a hormone known as melatonin at levels different in the light and in the dark, high in the dark and low in the light.

Information about light seems to get to the pineal gland in an unusual way. Most of the nerves coming from the eyes go into central information-processing stations that are involved in our ability to see. In addition to these main routes for information from the eyes, there are also two rather small bundles of nerve fibers that go into a region at the base of the brain known as the hypothalamus. This region is responsible for many of the body's basic life-support functions, such as eating, drinking, and temperature control. All the evidence available to date indicates that the hypothalamus is also the location of the body clock. And whatever else it does, it regulates the output of melatonin from the pineal gland.

A dose of melatonin makes people feel sleepy. Some recent experiments by Josephine Arendt and her colleagues at the University of Surrey suggest that melatonin may also help reset the daily rhythms of the body. For the experiment, a group of volunteers flew from London to San Francisco, where they spent two weeks adjusting to the new time zone. Three days before the return flight, some of the volunteers began taking melatonin at six in the evening while others received placebos, identical pills that contained no hormone. On their return the volunteers continued to take the pills, now at their bedtime, for four days. By all reports, the melatonin takers suffered significantly less from the effects of jet lag than did those who took the dummy pills. This still does not prove that the melatonin helped the travelers shift their clocks. That will require considerably more research. But it does suggest that there may be a means of alleviating the disruptive effects of hopping time zones.

Another hope for a jet lag pill comes from experiments with animals. This time Fred Turek and Susan Losee-Olson of Northwestern University used a tranquilizer to alter the sleep-waking pattern of golden hamsters. The tranquilizer was one of a variety known as a benzodiazepine, which is discussed in the section on mental illness and drugs. It is commonly used in the management of stress and insomnia. It also alters the action of brain cells found in the region of the hypothalamus that is associated with the biological clock.

In their experiment Turek and Losee-Olson observed the daily running patterns of hamsters kept in either constant light or constant darkness, then injecting them with the tranquilizer at different times before, during, or after their daily running period of a few hours. The results showed that the tranquilizer can either advance or delay the running session by as much as an hour, according to when it is given. If the dose is given about six hours before the start of a daily run, the hamster starts its next day about an hour earlier. If it is given about nine hours after the start, running the next day is delayed about an hour. These results differ from what happens when light changes are used to change the timing of the hamsters' clocks, which suggests that the

It Takes Timing: The Rhythms of Life

Dreams and How to Direct Them
Les rêves et les moyens de les diriger by Marquis d'Hervey de Saint Denis, 1867.

chemical has a different route of access to the clock mechanism than does the light. If these results hold up in tests with humans, they suggest that it may be possible to help the traveler reset his clock with a simple pill.

Knowing how our body clock is controlled and gaining the ability to reset that clock are obviously important for jet-setters, as well as for people who have no choice but to shift work schedules from time to time. There is another area of human endeavor in which our ability to get at the hands of the biological clock is going to be absolutely critical. That is in space travel. Leaving the surface of the earth in a space vehicle is very much like the laboratory situations referred to earlier in which cues about night and day are not available. But in space the problems are even worse, and disruption of biological rhythms seems to be one of the principal problems likely to afflict the space traveler.

Because the cycles of sleeping and waking are the most obvious indications of the running of our biological clocks, and the functions of sleep and associated dreaming have long puzzled mankind, sleep and dreaming have been intensively studied since ancient times. It is only within recent years, however, that scientific investigations have begun to shed any real light on the difficult questions of what is going on in the mind and the body during the particularly inaccessible behaviors of sleep and dreaming.

The behavioral characteristics of sleep in humans are quite obvious: the sleeper assumes a characteristic horizontal posture and becomes relatively motionless and unresponsive to external stimuli. The common notion is that this behavior serves to restore the brain and body after a day's activity, but the precise nature of this restorative function is only now being uncovered.

Sometimes upon awaking, a sleeper will report an extraordinary hallucinatory experience in which all manner of bizarre and sometimes frighteningly realistic events occurred, an experience called dreaming. The ancient Greeks regarded dreams as revelations from the gods and encouraged such revelations by sleeping in sacred places. Some centuries later Sigmund Freud confected a completely different theory of dreams, associating them with the suppressed infantile wishes and desires buried in his newly invented unconscious. It was not until the 1950s, however, that modern biological notions of the nature of dreams began to take shape, largely through the work of Nathaniel Kleitman and William Dement at the University of Chicago.

The modern tools for studying sleep and dreaming in humans are instruments for recording various sorts of electrical activity: the signals generated by the brain and recorded through wires applied to the scalp (the electroencephalogram), the electrical activity associated with muscle contractions (the electromyogram), and the electrical activity associated with eye movements (the electrooculogram). These tools, plus the verbal reports of sleepers, provide the dim insights we have into what happens in the mind during sleep.

There are two distinct states of normal sleep. The first is characterized by synchronous brain waves seen on the electroencephalogram, sporadic movements of the body, and rolling eye movements, which gradually subside. After an hour or so, the body muscles completely relax, and the electroencephalogram takes on the characteristics it has when the sleeper was awake, even though responses to outside stimuli are hardest to elicit at this stage. The electrooculogram then begins to show rapid movements as if the sleeper were following some action invisible to the observer. This last characteristic gives the label "rapid eye movement" (REM) to this sleep state. Observers have found that sleepers awakened during REM sleep usually report that they were experiencing the hallucinatory events associated with dreaming. In contrast, sleepers awakened during non-REM sleep report that rather routine thought processes seemed to have been taking place, if anything. An episode of REM sleep may last from 5 to 40 minutes and usually recurs every 90 minutes or so for a total of four or five times a night.

Research on animals suggests that we are essentially paralyzed during REM sleep, perhaps accounting for the feeling of being unable to move on awakening from a nightmare. Michael Chase of the University of California, Los Angeles, finds that the nerves that activate muscles in animals are electrically blocked during REM sleep. The highly emotional experience of dreams is reflected, however, in wild swings in heart and respiratory rate. At the same time, input to the brain is somehow blocked off. Nonetheless, the isolated brain appears to be highly active, appar-

It Takes Timing: The Rhythms of Life

1. Hibernation
Different animals display different behaviors controlled by biological clocks. Some, like many types of birds, migrate. Others, such as bears, marmosets, and dormice, hibernate.

2. In outer space
Beyond the pull of gravity, orientation becomes a problem. What happens to the brain in space?

It Takes Timing: The Rhythms of Life

Time lag
A sculpture by Arman at the Saint-Lazare train station in Paris.

ently generating the fantasies of the dream world from fragments of experiences stored within it.

Freud viewed dreams as having deep psychological import, revealing aspects of a person's mental life otherwise purposely kept hidden or even inaccessible to a person's own conscious thought. There is no evidence, however, that dreams reveal any more about the personality than do waking thoughts. More recently, Nobel laureate Francis Crick and others have begun to develop the notion that dreaming is a process of unlearning useless information, a kind of attic cleaning by the brain. Until we learn more of how sleep proves as refreshing as it does, we are left to guess. Is sleep necessary to pack information securely into our memory banks, to prevent mental breakdown from an overload of the useless information that roars through our sensory channels all day but lies ignored by our very choosy filter of attention? Until we know more about the biology of sleep and dreaming, dreams will remain as much the domain of the fortune teller as they are of the scientist.

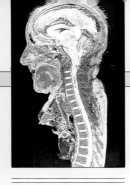

The Biological Clock

by Alberto Oliverio, director of the Institute of Psychobiology and Psychopharmacology, Rome

A great many biological functions follow a rhythmic pattern. In some cases the periods are extremely short, for example, in the rhythms of the bioelectrical events characteristic of cerebral activity. In other cases the periods can be up to a year or more.

The daily rhythms are synchronized with the movement of the earth on its axis (that is, with the alternation of day and night) and are present in many species of animals and plants. Many functions of organisms have a rhythm of this type, since the day-night cycle is extremely important for various aspects of the life of plants and animals, including human beings.

A daily rhythm in which rest and activity, sleep and wakefulness, alternate is clearly evident in the higher organisms and, in particular, in mammals. In the natural environment, the alternation of light and darkness in the course of a day is the main factor in synchronization. Species such as man are active during the daylight hours (although this is not true to exactly the same extent for all individuals), while the opposite is true of such nocturnal species as cats and mice.

Further examples of daily rhythms are provided by temperature and by the activity of the adrenal glands, which secrete corticosteroids, the hormones involved in stress, defense, and the maintenance of blood pressure. The cells in every tissue of our bodies, with the exception of those in the brain, have a daily rhythm based on the mitotic division of cells, the process by which the cells that die are gradually replaced. For example, in the case of the skin, this division takes place by night and reaches a maximum at around midnight.

Many physiological functions have a period of 24 hours even in the absence of the synchronizing factor of light and dark. In such cases we no longer speak of a daily rhythm but of a circadian rhythm. This indicates a rhythm with a free running period, i.e., not influenced by external agents. Such rhythms proceed in parallel with the rotation of the earth on its axis but do not coincide with it. If a rodent, such as a mouse or rat, that has previously been synchronized to a rhythm of 24 hours a day is kept in conditions of constant light or dark, it will develop a rhythm of sleep and wakefulness of 24 hours and 16 minutes (with variations of 2 minutes). This condition will be precisely maintained for months on end.

Remarkable progress has recently been made in our knowledge of the anatomical bases of the circadian rhythm in mammals and in analyses of pacemakers at the cellular and biochemical levels. The study of the neurophysiological correlates of circadian systems in mammals has allowed us to pinpoint the location of the pacemaker in the suprachiasmatic nucleus. The latter forms part of the hypothalamus and a portion of it serves to control recurring catering activities like sleep, wakefulness, hunger, thirst, and the control of body temperature. In the last decade, it has been established that the suprachiasmatic nuclei control many rhythmical functions in both man and animals, for example, spinning a wheel (an activity typical of many rodents kept in cages), drinking, eating, sleep, temperature, the levels of adrenocortical hormones (corticosterone), ovulation, the period of heat, and so on.

In man there are great differences of temperament, which partly reflect physiological differences. Some of these have been attributed to different roles of the two autonomous branches of the nervous system. People with a dominant sympathetic nervous system are supposed to be early morning types, to achieve better psychophysical performances during hours of daylight and, becoming tired sooner, to go to bed earlier. Those in whom the parasympathetic nervous system dominates are supposed to be evening types, that is, they have a tendency to rise later, to be tired in the morning, to achieve their best psychophysical performance in the evening hours, and to stay up late without tiring.

There are, in fact, certain physiological differences between these two categories of people (for example, the evening types tend to have lower blood pressure in the morning), some of which are circadian. One of these is connected with the bodily temperature during the day. For the early morning types, temperature reaches its highest peak roughly 70 to 80 minutes before that of the evening types. Other cyclical fluctuations observable in man are related to the levels of adrenal hormones and ACTH (the hypophyseal hormone that stimulates the adrenal cortex). The peaks of ACTH activity, frequently coinciding with meals during the hours of daylight, result in an increase in psychomotor performance, which reaches its highest level for sportsmen in the early hours of the afternoon. Other activities, such as sexual activity, also have a rhythmical character. For example, the level of male sexual hormones is higher during the night hours, and it is then that the sexual drive reaches its peak.

Several human activities have been studied in the absence of factors that can indicate the passing of time and alternation of day and night to subjects. In some cases, the subjects were kept in bunker-like rooms isolated from external light and sound. They were provided with books and record players for their entertainment and were free to switch the light on or off at will. Other experiments studied the reactions of speleologists who remained underground in caves for weeks or months at a time. The results of these experiments show that many functions and patterns of behavior retain a rhythmical cycle in these conditions.

Many studies have been carried out on human beings to observe the effects of phase shifts, a frequent phenomenon nowadays for those who travel by air between countries with marked time differences. For example, a passenger going from New York to Rome leaves New York at four in the afternoon and arrives in Rome eight hours later. According to New York time it is midnight, but according to Rome time it is six o'clock in the morning, and he has a new day to face with a new rhythm. Those traveling from east to west have to adjust their biological clocks to a longer day because of the time gained by traveling toward the west. The problems caused by jet lag have been the subject of a great deal of research on the effects of a twelve-hour shift. Subjects used to a certain rhythm (e.g., going to bed at ten in the evening and waking at six in the morning) were made to reverse the order of these activities by going to bed at ten in the morning and waking at six in the evening. Certain physiological parameters quickly adapt (cardiac rhythm, for example), while others, like bodily temperature, the adrenal hormones secreted in conditions of stress and some electrolytes (sodium, potassium) take an average of six to seven days to catch up with the new rhythm. There are considerable differences between individuals, some adapting very quickly and others requiring days. Tests show that psychomotor performance also deteriorates rapidly after the inversion of rhythm (largely through lack of sleep) and takes at least six to seven days to return to a normal level.

One of the developments of research into circadian rhythms is related to the study of the effects of drugs at different times of the day. Not only those affecting the nervous system but also those used for the heart, kidneys, and tumors show greater or lesser activity according to the time of day. This knowledge enables us to use antineoplastic drugs in smaller, less toxic doses.

The study of drugs in relation to circadian rhythms (chronopharmacology) is a recent, important development, especially with regard to drugs affecting the nervous system. The effects produced differ according to the time of day at which they are administered. A typical case is that of antidepressants, which are more active in the morning than in the late afternoon. These biphasic effects seem to be connected with fluctuations in the receptors upon which the drugs act. The same fluctuations are responsible for the change of mood that takes place in most people during the hours around sunset, when there is a tendency toward greater anxiety or depression. These studies also have useful therapeutic implications, since they allow us to clarify certain aspects of depression.

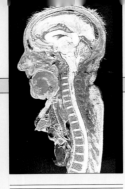

Sleep and Dreaming

by Allan Hobson, Harvard University

The brain awake and asleep

The brain operates differently during the waking state than it does during the sleeping state. The instrument that best reveals these differences, the electroencephalograph, detects, amplifies, and records electrical signals given off by the brain through electrodes placed on the scalp. A record of these signals, or brain waves, is referred to as an EEG (electroencephalogram).

In the awake state, the brain is activated and the EEG is characterized by fast-frequency, low-amplitude brain waves, the so-called alpha rhythm. During sleep, however, the alpha rhythm disappears and other brain-wave rhythms are generated. Not only does brain electrical activity change during sleep, but other, important physiologic changes also take place. Studies of sleep will usually include all-night monitoring of such changes, in addition to monitoring brain electrical activity. Eye movements (EOG), muscle tone (EMG), and cardiac activity (EKG) are among the standard physiologic variables measured during sleep.

Perchance to dream

In the early 1950s a discovery was made that significantly changed the course of sleep and dream research. The discovery was that at regular intervals during sleep, the eyes of a person rapidly move back and forth beneath the eyelids. Brain electrical activity associated with these rapid eye movements was strikingly similar to that of an awake EEG, yet the subjects in whom this phenomenon was observed were clearly asleep.

It was subsequently learned that if a subject was awakened during one of these periods of rapid eye movements (REM), he frequently reported he had been dreaming. Often he would give long, clear, detailed accounts of his dream.

The discovery of REM thus revealed that there were two kinds of sleep that alternate rhythmically with each other throughout the night. One kind of sleep, now referred to as REM, is the sleep during which most dreaming takes place.

Because of the paradoxical similarity between REM sleep and awake EEGs, scientists began to examine other physiologic variables during REM to determine if other changes might also be taking place. They found that in addition to an activated EEG, another striking characteristic of REM was a generalized motor inhibition, which was manifested by an absence of muscle tone in the EMG. Visual observations of subjects during REM showed there were twitching of facial muscles and fingertips, irregular respirations, cessation of snoring if present prior to the onset of a REM episode, and penile erections in males. The latter, incidentally, is a physiologic concomitant of REM and is not related to dream content, as one might suspect.

The other kind of sleep that alternates with REM is called non-REM. It is distinctly different from REM sleep and is usually described in terms of its four stages, defined by specific EEG characteristics. Stage 1 does not last long and sleepers may be easily awakened during this stage of non-REM sleep.

Over the course of the next half hour or so, sleep becomes deeper as stages 2, 3, and 4 of non-REM progress. The EEG of stage 2 shows bursts of brain wave activity resembling spindles. The eyes continue to roll and the muscles continue to relax. Arousal is more difficult in stage 2 than in stage 1. Stage 2 sleep gives way to an even deeper sleep, stage 3, in which the brain waves are larger and

The rest of the night

slower. At this point the sleeper cannot be awakened easily; his heart rate and respiration become slow and very regular.

The last stage of non-REM is stage 4, the deepest sleep. Brain waves are large, slow, and regular. The body muscles are so relaxed that few movements are likely to be observed. After several minutes of stage 4 non-REM sleep, a sleeper usually reverts for a while to one of the lighter stages of non-REM sleep.

When roughly 90 minutes have elapsed, the first installment of REM appears. Its duration is brief and it is followed by one of the non-REM stages of sleep. This pattern of alternating REM and non-REM sleep repeats itself four to six times a night, depending on the length of the whole sleep period. Later in the night, successive REM periods become longer and less time is given to the deeper non-REM stages. For example, the first 90-minute cycle of sleep might consist of 85 minutes of non-REM and 5 minutes of REM; the fourth 90-minute cycle of sleep might consist of 60 minute of non-REM and 30 minutes of REM. Thus, the likelihood of being awakened during a dream increases toward the end of the sleep period. This is why we tend to remember those dreams that occur close to the time of being awakened.

Incidentally, dreaming is not limited exclusively to REM sleep: subjects awakened from non-REM sleep are likely to report having mental activity of a thoughtlike nature. For example, someone awakened during non-REM might say, "I was thinking about getting my car fixed tomorrow." This is clearly different from that kind of report elicited during a REM awakening, but it also shows that there is mental activity during non-REM sleep.

Another view of sleep

The cyclic nature of sleep, as revealed by the human EEG, strongly suggests that it may be under the control of a master biological clock that governs the 24-hour alternation between waking and sleeping characteristic of all mammals. REM sleep, then, may be regarded as a mini-rhythm within the 24-hour circadian cycle. Behavioral studies in which sleeping subjects were photographed at fixed intervals throughout the night support this notion. Body movements are not seen in the photographs that correspond to stages 3 and 4 of non-REM sleep. A major postural change may occur, however, during the transition to REM and again at the termination of REM, when the non-REM phase of the sleep cycle is resumed.

It is interesting to note that this behavior is not exclusively limited to sleeping human subjects. Cats similarly photographed also show these postural adjustments at intervals corresponding to their cycle of non-REM and REM sleep. This finding and the similarities in cat and human brain electrical activity tend to make the cat an ideal animal model for studying sleep and dream phenomena.

Even before the discovery of REM it was known that neurons in a part of the brain stem called the reticular formation were somehow involved in the control of sleeping and waking. It was also known that electrical stimulation of certain neurons in the reticular formation produced a generalized inhibition of motor movements. Following the discovery of REM sleep, characterized by EEG activation and muscle atonia, a series of lesion experiments revealed that cells in the pontine brain stem were responsible for the maintenance of REM sleep: selective lesioning of pontine cells actually eliminated REM sleep in the cat. If pontine cells were left intact and all other parts of the brain above the pontine reticular formation were destroyed, REM sleep was preserved.

Another group of neurons in a nearby part of the brain stem, the locus ceruleus, was also found to be important, since its destruction could eliminate the muscle atonia of REM

Brain stem neurons and REM

sleep but not its other manifestations. This meant that REM sleep might depend upon an interaction of cell groups in the brain stem. Since the neurons in the locus ceruleus had an additional property of being chemically coded with the inhibitory transmitter norepinephrine, a possible chemical element in REM-sleep regulation was suggested.

Giant cells: The REM generators

Today there is abundant evidence pointing to a group of giant cells in the pontine reticular formation of the brainstem as the neurons most likely to be responsible for the generation of REM sleep. Immediately prior to and during each REM episode the neural activity of these giant cells greatly increases. At all other stages of the sleep-wake cycle their neural activity is minimal or nonexistent because of inhibitory nerve impulses coming from a nearby group of neurons in the locus ceruleus. Periodically, however, inhibition by the locus ceruleus neurons ceases and the giant cells become active. The result is REM sleep.

During the brief periods of unsuppressed activity, the giant pontine cells send excitatory impulses to many areas of the brain, particularly the cerebral cortex, by way of their widespread fiber projections. The content of a dream may then be synthesized in the areas of the cortex thus activated.

Rapid eye movements and the visual images of dreams

A striking feature of dreams is their rapidly shifting, intense, visual nature. This feature has been thought to be somehow associated with the rapid eye movements that characterize dreaming sleep. Some scientists have postulated that the eye movements represent the dreamer scanning the visual imagery of his dream.

An alternative explanation is that eye movements of dreaming sleep precede the visual events of the dream: neurons that regulate eye movements become activated by the giant pontine cells, the REM generators. At the same time, the cerebral cortex receives neural information about the speed and direction of the eye movements and utilizes this information to synthesize an image. A specific example may help to clarify this concept. Suppose you are dreaming and in your dream you see the back of a man who is standing at an intersection. Suddenly the man turns to the left and runs across the street. The explanation for this event in the dream would be that REM-generating pontine cells activated nearby eye-movement neurons, specifically those that move the eyes to the left. The cerebral cortex registered this activity and attempted to make sense out of it in light of what previously occurred in the dream. The logical solution, based on the speed and direction of the eye movements, was to move the man to the left and rapidly across the street. Unpredictable and sporadic eye movements may account for the unusual visual qualities of the dream. Their occurrence in bursts may, in fact, determine dream plots or sequences of action.

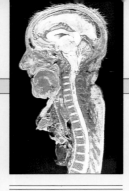

Alterations in Brain Function during Weightlessness

by Laurence R. Young,
Man-Vehicle Laboratory,
Massachusetts Institute
of Technology

During the weightless condition of free fall in an orbiting spacecraft, the brain no longer receives its usual pattern of consistent signals from the sense organs associated with spatial orientation. Although the semicircular canals of the vestibular system continue to correctly indicate angular velocity of the head, and the visual system signals the orientation of the head with respect to the floor, walls, or other dominant lines, the gravity-sensitive organs respond in a totally unaccustomed fashion. In particular, the otolith organs of the inner ear, which normally behave like a balance or pendulum to indicate the direction of the head with respect to the vertical, lose this function entirely. Just like a pendulum in weightlessness, the otolith-organ response is unrelated to head orientation and depends upon random minute forces on the organ and on where it was left at the end of a previous movement. Indeed, the otolith organs within a single ear and between the two ears are likely to disagree in their indication of down. Far from being deafferented or useless, the otolith organs continue to function as linear accelerometers and supply vital information to the brain concerning the linear acceleration and deceleration that an astronaut undergoes as he moves about the spacecraft or as the spacecraft itself is accelerated by orbital maneuvering thrusts.

Thus, the central nervous system's interpretation of sensory signals from the otolith organs ought and indeed appears to change in weightlessness. Otolith outputs may be reinterpreted in terms of head acceleration rather than tilt, according to this theory.[1]

Direct evidence concerning adaptation of the central nervous system to new ways of processing spatial orientation information in weightlessness is still quite limited and some of it is anecdotal. Among the earliest was the report by some of the astronauts of an inversion illusion. Following entry into orbit, some crew members would feel that they and the spacecraft were upside down with respect to an unseen outside reference. An inverted view of the earth might contribute to this illusion but was not required. One explanation for this illusion is based upon the sudden change in gravitoinertial acceleration sensed by the otolith organs and all other inertial force sensors in the body. Prior to entry into orbit, the net force corresponded to an acceleration of greater than one g, which was suddenly reduced to zero when the rocket thrusters were turned off. This sudden removal of a bias greater than one g results in a change in acceleration along the longitudinal axis of the head similar to that experienced when standing inverted in a one-g field and might be interpreted in this manner. Another phenomenon that could account for or supplement the sensory explanation is the fluid shift toward the head. When going into weightlessness and for some hours thereafter, a significant shift of fluid from the legs and lower thorax to the upper part of the body occurs, resulting in a fullness in the face, stuffy nose and sinuses, and other sensations normally associated with standing inverted. For some astronauts, this inversion illusion does not appear immediately but only after several hours in orbit, and it may last for a period of hours to days. The inversion illusion is frequently disturbing and associated with symptoms of a malady known as space motion sickness.

"Space motion sickness" describes a collection of signs and symptoms (closely related to seasickness). It affects approximately half of all space travelers. It normally begins shortly after entry into orbit and is usually over within three days or less. Nausea, vomiting, loss of appetite, and changes in peripheral circulation are among the symptoms. Space motion sickness is frequently exacerbated by any activities that present a further sensory conflict between the response of the otolith organs and the other sensors. Pitching or rolling head movements in particular are frequently provocative. These head movements present a strong conflict between the semicircular canal signals indicating rotation and the otolith organs that fail to confirm a change in head orientation. Visual signals, if present, and proprioceptive signals, of course, agree with the semicircular canal information. Similarly, unusual visual cues, such as seeing another crew member upside down emerging into the spacecraft in an unusual position or seeing the world inverted through a window, may all contribute to sensory conflict and consequently to space motion sickness.[2] Some drugs have been used with partial success to manage these symptoms (especially scopolamine with or without amphetamine and a combination of promethazine and ephedrine), but none has been shown to prevent the symptoms. The crew members who do not experience space motion sickness are not necessarily those who are insusceptible to seasickness or to motion sickness during various tests on the ground. Functional treatment of symptoms includes minimizing the conflict by restricting head movements and possibly closing the eyes and generating unambiguous tactile cues of orientation by wedging into a corner or applying forces to the feet through stretched elastic cords attached to a shoulder harness.

The disappearance of spatial orientation illusions and symptoms of space motion sickness over the first few days in orbit suggests that adaptation by the central nervous system might include a reinterpretation of otolith signals and increased weighting of other sensory information, including tactile and visual information.[3] Tests of responses to uniformly moving wide visual fields that produce a sensation of self-motion known as vection support these findings. Circular vection and linear vection are stronger in weightlessness than on earth. This can be interpreted as evidence that the nervous system places increasing weight on visual information and learns to ignore the no-longer-appropriate otolith information that might otherwise conflict with and inhibit the visual responses. Compensatory eye movements, which also respond to visual and vestibular stimuli, may also show evidence of changes in the weighting of these cues in orbit, although results to date are limited. The commonly observed asymmetry in vertical nystagmic eye movements (rhythmical oscillations) in following a large moving field is apparently altered in weightlessness, when the steady gravity bias on the otolith organs is no longer present.[4] Postural responses to linear acceleration toward the feet, normally associated with falling, also appeared to be reduced in weightlessness.[5]

In general then, the brain appears to undergo a measure of appropriate adaptive responses to the altered sensory inputs it receives in weightlessness. This adaptation, which may involve changing sensory weights, is appropriate for living in zero gravity but leads to a number of inappropriate sensations shortly after return to earth. Postflight postural instability and illusions may last from several hours to several days after return.

6
Emotions: A Gut Feeling

Emotions can be troublesome not only in everyday life but also to scientists trying to figure out the relationships among emotions, mind, and brain. Except for unfortunate patients with certain forms of brain damage, we all experience emotions, both positive and negative. Some individuals feel that emotions enhance the experiences of life. Others find them a disturbing intrusion into rationality, annoying evolutionary remnants from our primitive ancestors. After all, emotions are generally associated with uncontrollable alterations in the internal organs, particularly the heart and the gut. The discovery some years ago that emotions are largely mediated by a part of the brain, the limbic system, that arose early in the course of evolution confirmed the suspicions of those who hold emotions to be unwanted intruders into the lives of humans.

One of the big problems with emotions is the difficulty of controlling them. Children must learn how to do so. Many cultures insist that no matter how intense an emotion might be, it should never be displayed. How is an experience so intense yet so elusive as love to be scientifically analyzed and dissected? A Shakespearean sonnet may evoke in the mind a resonance of the feelings of love that stir the heart, yet it cannot cause those same feelings. But the idea of resonance is an appealing one. Somehow the mind must register the gut-stirring experiences that seem to be essential components of emotion.

That emotions hold a special place in our mental lives is apparent from examinations of patients with brain damage. Some strokes, for example, may cause a peculiar condition known as prosopagnosia, in which the patient is unable to recognize faces, even those as familiar as a spouse's or a child's. Even though the patient is unable to identify whose face he sees, if it is the face of someone emotionally close his heartbeat will nonetheless accelerate. Thus the visual stimulus can evoke an emotional association even when the verbal association is lost.

The contemporary view of how mind and visceral activity are linked owes much to psychologist Stanley Schachter of Columbia University. He developed the idea that visceral arousal is necessary for emotional experience, but that the nature of the emotional experience will depend upon an individual's thoughts, memories, and current circumstances. George Mandler of the University of California, San Diego, characterizes this view in the following terms: "Riding on a roller coaster produces serious disturbances in discrepancies between our expectations and current feelings of balance and bodily support. Whether the ride is seen as joyful or dreadful depends on what we expect about the ride, who accompanies us, what we are told to expect, and whether we feel in control of the situation. Some love it, others hate it."[1]

We do not know very much at all about the neurological organization of human emotion, that is, how the brain is organized to integrate mental processes with visceral events. There is only circumstantial evidence that, for example, emotional events are remembered especially well because parts of the visceral brain are also involved in memory formation. There is considerable evidence from brain-damaged patients that the right hemisphere of the brain is responsible for the expression and perception of emotional states. Stroke victims with right-hemisphere damage may, for example, be unable to modulate their voices to express emotion and may fail to perceive emotional aspects of facial expression. But these generalities tell us little about the actual events underlying the experiences of emotion.

We can make some associations between what we experience as emotions and how animals, whose brains we can examine, behave in certain emotional situations that trigger actions we might associate with such emotions as fear or rage. There is, of course, some risk in trying to extrapolate from what we know about seemingly emotional behav-

The psychology of crowds
The psychology of crowds, which has its roots in Freud's Civilization and Its Discontents *(1930), was developed by the French journalist Gustave Le Bon (1841-1931). According to Le Bon, "By the mere fact that he forms part of an organized crowd, a man descends several rungs in the ladder of civilization: in a crowd, he is a barbarian that is, a creature acting by instinct."*

ior in animals to emotional behavior in humans. As José Delgado, Spanish neurophysiologist and expert on aggressive behavior, has noted, "Mouse-killing behavior in the cat has very little in common with organized crime."[2]

Brain research on emotions in animals has led to some understanding of the parts of the brain that are involved in such primitive, reflexlike emotions as rage and fear. Those emotions thought to be solely in the realm of human experience, including love and sorrow, remain beyond the grasp of the neuroscientist. Only when stroke or other brain damage disrupts someone's ability to experience such emotions do we have the reassurance that these emotions too are a function of brain cells and not some elusive attribute of nonbiological spirit.

We assume that emotional behavior in animals, like all behavior, must have developed because it has, or had, some value for the animal's survival. One emotional behavior that is receiving considerable attention today is a reaction commonly known as the stress response. It was originally characterized as the fight or flight response. It occurs, for example, when one animal threatens another. The threatened animal may respond with an enraged counterattack, or it may turn tail and flee, according to the nature of the threat. These behaviors have clear survival value. Most people can clearly recall the emotional components of this response when responding with rage to a playmate's senselessly throwing a stone or when fleeing in terror at the approach of the neighborhood bully.

One of the oldest ways to study the involvement of various parts of the animal brain in different behaviors is to insert a fine wire into the brain region of interest. A minute electric current delivered through the wire will activate the cells in that region and possibly elicit some specific behavior. Some decades ago, brain researchers discovered that they could cause animals to display ragelike behavior by stimulating various brain regions, especially in the limbic system and the hypothalamus. This discovery led to the conclusion that there are certain circuits of brain cells analogous to electrical circuits that give rise to rage behavior. There is still some reason to believe in the existence of these kinds of circuits, but we know now that they are not as inflexible as researchers once believed.

The flexibility of brain circuits can be seen in monkeys that form colonies in which there is a hierarchy of dominance among the animals. The top animal dominates all the others. The one below him dominates all except the top monkey, and so on down the ranks. Electrical stimulation to an appropriate region of a dominant animal's brain will elicit a rage response, but stimulation of the same region of the brain of an animal low on the social ladder elicits submissive behavior. In humans there are probably a great many factors that have an equally powerful effect in modulating emotions, many of which have to do with cultural learning and have no counterpart in animals.

Let us return now to the stress response. Current research in this area focuses on hormones in the brain whose release triggers many of the familiar changes that occur in the body in threatening situations. These include an increase in heart rate and blood pressure, and the peculiar feeling associated with the rush of adrenaline into the bloodstream. In recent years all of these responses that prepare the body for intense physical activity have been tied to the release in the brain of the hormone corticotropin-releasing factor (CRF), first identified by Wylie Vale at the Salk Institute. Cells in the hypothalamus send CRF to the pituitary gland, where it stimulates the release of another hormone, adrenocorticotropin. This hormone flows through the bloodstream to the adrenal glands perched atop the kidneys, where it causes the release of the steroid hormone cortisol. At the same time that this chain of hormonal events is occurring, CRF in the brain activates brain cells whose signals lead to an increase in heart rate and the release of adrenaline.

There is a dark side to stress because it presumably evolved to aid survival in the face of physical threats but is now inappropriate for the emotional stresses of modern humans. Psychiatrist Philip Gold and his colleagues at the National Institutes of Mental Health have been studying patients with melancholic depression and find that they display all the signs of being in a state of perpetual stress. In animals it appears that the parts of the brain involved in such emotions as anxiety, normally associated with stress, can become sensitized to CRF and become activated with little provocation. Furthermore, long exposure to cortisol can destroy brain regions that normally shut down the stress response when they detect the presence of cortisol, allowing the response to continue unchecked. This picture remains to be verified in humans, but holds out hope for the treatment of severe emotional disorders.

A major question yet to be answered is how the perception of a threatening situation, whether real or imagined, activates the critical CRF-containing cells in the hypothalamus. New techniques for tracing brain-cell connections are revealing that the CRF-containing cells receive signals from cells in the limbic system, which is so strongly involved in emotional experiences. But if there is any one seat of emotions, it may well be the hypothalamus. As noted previously, cells in the hypothalamus are involved in such behaviors as eating and reproduction, which have very strong emotional components, as well as in rage and fear.

Furthermore, some years ago, scientists found that electrical activation of a certain region of the hypothalamus seems to evoke sensations of pleasure in animals. They will exhaust themselves pressing a lever that causes electrical stimulation of this region. The idea that such stimulation provides pleasure was later confirmed when in the course of exploration for epileptic foci, human patients received electrical stimulation in this region and reported pleasurable sensations. So while there is no generally accepted theory of the brain's role in emotional experiences to guide research in this area, there are at least the beginnings of ideas about how our emotional experiences are related to the organization and activity of brain cells.

1. Ronald Reagan
"One must laugh before being happy, lest we died without having laughed," wrote the French moralist Jean de La Bruyère.

2. Facial expressions
The Mechanisms of Human Physiognomy, or the Electrophysiological Analysis of the Expression of Feelings, *a work in French by G. B. Duchenne de Boulogne.*

7
Strange Thoughts: The Disordered Mind

Disorders of the mind provide some of the most fascinating glimpses that brain researchers can gain into the interplay of mind and brain. When something goes wrong with a person's ability to think or to feel emotions, the resulting abnormal behavior reveals how delicately balanced our brains actually are. We often feel uneasy about this, but we also begin to realize how little we really know about the connections between the brain and the mind.

Throughout this book we see growing evidence that our impressions of what our minds are doing are very poor guides to understanding what is actually going on in our brains. So we must be even more cautious when considering brain functions associated with disorders of the mind. The unfortunate sufferers of disorders are even less reliable reporters of what they think their brains are doing than persons whose minds are in more or less normal condition. And there are many ways for observers, even trained professionals, to make up stories to explain aberrant thoughts or moods. Great bodies of conflicting psychoanalytic theory testify to the difficulty of discovering in this way what is actually wrong with minds that are not working properly.

Not surprisingly, there are many arguments among professionals about the treatment of mental disorders. We know so little about the real nature of the causes of schizophrenia, depression, phobias, and other disorders that truly effective treatments are often not available. Some critics claim that there is, in fact, no such thing as mental illness and that persons whom society labels as mentally ill simply suffer from societal stigmas. The confusion over treatments for mental disorders may be even greater than the apparent confusion over what the disorders are and how to classify them. Do drugs work? What about the dozens of offshoots of Freudian psychoanalysis, behavior therapy, and humanistic therapies?

There are no quick and simple answers to all these questions, and here more than anywhere else in research related to the function of the brain it is abundantly clear that we are not yet clever enough or wise enough to ask the right kinds of questions. Questions about mental disorders are motivated by observations of behavior that is in some sense abnormal. But questions about the definition of "abnormal" lead to murky areas of societal judgments and ethics, not to scientific explanations of brain function. Perhaps a better kind of question would be to ask what we know about the biological causes of unusual patterns of thinking or acting that somehow make people out of touch with what most people consider to be reality.

Schizophrenia is undoubtedly the most widely recognized group of mental disturbances. One person in every hundred is likely to be afflicted sometime during life, regardless of country or culture. The word "schizophrenia" comes from Greek words meaning split mind, and many people often confuse it with the much rarer behavior disorder known as multiple personality. Schizophrenia is really a fragmentation of just one personality, whose pieces can take on many different aspects. Thinking can become disjointed, with the patient unable to maintain a continuous line of thought. It becomes impossible for some schizophrenics to pay attention to what is going on around them; they become constantly distracted by what is going on inside their own heads. They may become socially withdrawn, lose interest in their surroundings, and suffer hallucinations. Such hallucinations are usually in the form of voices that discuss the patient as "he" or "she," an unpleasant sort of eavesdropping. This is only a small sample of the effects associated with schizophrenia, but it is enough to make clear the fact that something quite real is happening as far as the patient is concerned.

Is schizophrenia a mysterious alteration of thought and behavior patterns, or does it

Voodoo
Voodoo is a ritualistic practice derived from African ancestor worship and found today principally in Haiti. Participants in voodoo rites make animal sacrifices and supposedly experience trances and other altered states of consciousness. Practitioners of voodoo claim to be able to revive dead bodies as zombies, the "living dead." Some students of voodoo claim that zombies are simply people drugged with small doses of the poison from blowfish, which sometimes kills Japanese fond of this delicacy. This assertion has not been conclusively proved, however. Zombies remain a favorite subject of horror movies.

have some biological origins in the brain? Despite years of searching through the chemical contents of the blood, urine, and brains of schizophrenic patients, scientists still do not have a definitive answer. But the fact that there is a clear hereditary aspect to schizophrenia suggests that there is some underlying biological component of the disorder. Even more compelling evidence that schizophrenia may involve abnormal brain function comes from the effects of drugs that counteract the effects of schizophrenia. Chlorpromazine and related drugs that came into use in the 1950s created a revolution in psychiatry and introduced the idea of biological psychiatry. The possibility of treating mental disorders like diseases of the body became real for the first time.

Did this mark a new beginning in understanding the mind in biological terms? Despite the initial excitement and a few false alarms since, it regrettably did not. We have a few hints about how chlorpromazine might affect the brains of schizophrenics, but the knowledge is not definitive and there are several competing theories that have not been disproved. One theory, for example, says that a basic difference in the body chemistry of schizophrenics causes them to convert some normal food substance into a kind of drug, like mescaline perhaps, that causes hallucinations. No one, however, has been able to pin down any such chemical tricks.

Another hypothesis about the chemical nature of schizophrenia is bolstered by a few more facts. It relates to the actual effects on the brain of substances that can improve a schizophrenic's symptoms, like chlorpromazine. Chlorpromazine and related substances interfere with a chemical system that carries messages between brain cells. One cell releases a neurotransmitter into the minute space between it and a neighboring cell. When the neurotransmitter arrives at the neighboring cell, it attaches to a receptor that is precisely shaped to receive the neurotransmitter, much as a lock receives a key. Once the neurotransmitter arrives at the receptor, it can either tend to activate the receiving cell or to prevent it from being activated by other transmitters, according to whether the first transmitter is excitatory or inhibitory. (This process is discussed in more detail in chapter 12.)

In the case of schizophrenia, the neurotransmitter that may be involved is known as dopamine. The hypothesis is that somehow the dopamine system in one or more parts of the brain of the schizophrenic is overactive. Chlorpromazine interferes with the interaction between the dopamine and its receptors, and this interference may be what helps reduce the severity of the symptoms. The validity of this proposal is further supported by the fact that drugs that increase the activity of dopamine systems in the brain, like amphetamine, produce psychotic episodes. In some ways these episodes resemble schizophrenic symptoms. Other strongly suggestive evidence was recently found using positron emission tomography, a technique that can produce images of the distribution of specific chemicals within the living brain. A team of researchers at the Johns Hopkins University observed that a group of schizophrenic patients had a higher density of dopamine receptors in certain parts of their brains than did a group of normal volunteers. Whether this increase is specific to schizophrenia or also occurs in other psychotic disorders remains to be determined.

An important consideration in the story of both helpful and harmful drugs that affect the brain is the molecular "security screen" known as the blood-brain barrier. Since the turn of the century it was known that the entry of chemicals into the brain is a privileged process; only certain substances can get in or out. It makes sense since a slight chemical imbalance in the brain can disrupt its operations and endanger life. The presence of the security screen was demonstrated early on by injecting dye into the bloodstream of experimental animals and showing that despite the fact that the dye stains all organs of the body, it does not stain the brain. It was not until the 1960s, however, that electron microscopy could demonstrate that the blood-brain barrier is formed by the cells that make up the walls of capillaries in the brain and the tight junctions these cells form around the passageways for blood flow.

Chemicals get across the blood-brain barrier in just a few ways. Some cross because they can dissolve in the fatty molecules that make up the walls of the barrier cells. This means that oily substances can get in and out readily, but even simple things that do not dissolve in oil, like sodium or potassium, cannot cross freely. Essential nutrients like glucose do not dissolve in oil, and so they must rely on

Strange Thoughts: The Disordered Mind

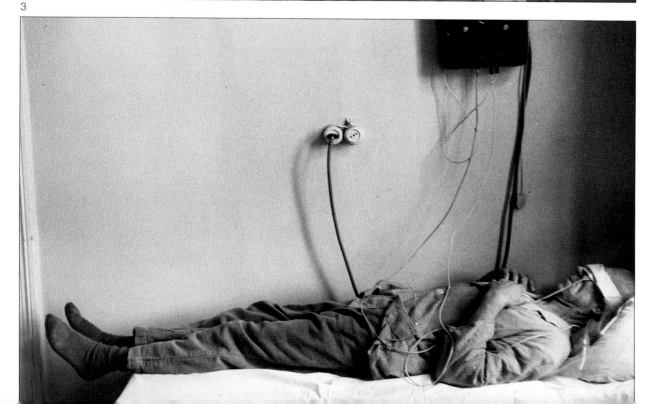

1. A Cat
A drawing by a mental patient.

2. Open psychiatry
In a small French village in Burgundy, the mentally ill of an open psychiatric clinic participate in the annual festival celebrating its patron saint, Alise.

3. Mental illness in Russia
A patient in the famous Kaschenko hospital, the Bellevue of Moscow.

Strange Thoughts: The Disordered Mind

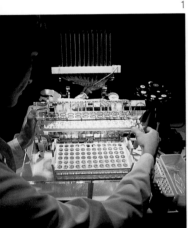

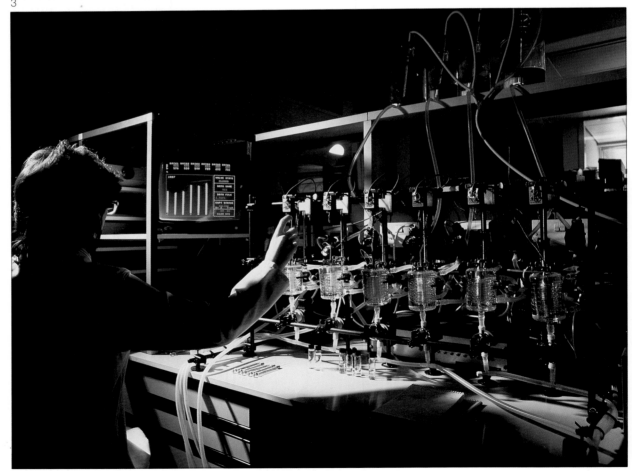

1. Receptors in the brain
Researchers use this equipment to study the biochemistry of receptor molecules in the brain that receive chemical messages from brain cells. They isolate receptors from brain cells and attach radioactive labels to them. They then measure how strongly the receptors attach to various drugs, thus determining the likely strength of different drugs.

2. The distribution of receptors in the brain
Radioactive labels on molecules that attach to different receptors make it possible to map the location of the receptors in the brain, as seen on the computer screen.

3. Pharmacology and the computer
Responses to drugs or natural chemical messages from the brain are displayed on a computer screen.

costly, energy-dependent transport systems to get into the brain.

Apart from the easy route of drug-induced altered states of consciousness, there appear to be other ways to interfere with a person's awareness of what is going on in the world. One is hypnotism, a phenomenon that has received little attention from brain researchers. Another is transcendental meditation. This is a mental phenomenon that goes on unobserved inside the head. It is not much of a behavioral phenomenon and is therefore difficult to study in the laboratory. But there have been a few attempts.

Meditation is a challenge to scientific ways of thinking in the same way that religion is. If meditation is in the same class as religious miracles and therefore by definition beyond scientific explanation, there is no point in discussing it here. Quite a few brain scientists, however, have tried to find out what happens to the brains of "enlightened" meditators. Some say that the meditators just go to sleep. Others are not so certain. The first problem, of course, is finding someone who truly possesses the ability to enter the advertised state.

One researcher who has made a serious effort is Peter Fenwick of Saint Thomas Hospital in London. He is a cautious fellow and for that reason worth listening to, despite much nonsense that has been put into print about the biological effects of meditation. According to reports, Fenwick has studied at least one self-professed Zen master. It appears from recordings of its electrical activity that the Buddhist's brain shifts its activity into high gear in the right hemisphere. Apparently he is not just going to sleep. This right-hemisphere shift seems to happen with other dedicated meditators as well, although, as Fenwick allows, it may have as much to do with the brain function of those persons willing to stick out the boredom of meditation as it does with the effects of meditation itself. But as our discussions here make clear, we still cannot make much sense of how the brain relates to consciousness, so it is likely to be quite some time before a book on brain research can say anything meaningful about meditation and inner peace.

Before we turn to matters connected directly to the structure of the brain and the details of how its components function, now is a good time to look both ahead to the connections between mind and brain that come

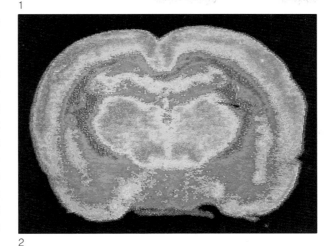

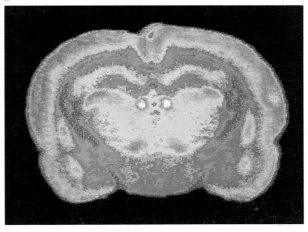

Antipsychotic drugs
These figures illustrate the effects of an antipsychotic drug on brain metabolism in a laboratory rat. Brain cells consume energy mostly in the form of the sugar glucose transported by the bloodstream. The more active a brain region, the more glucose it consumes. Louis Sokoloff and his colleagues at the National Institutes of Health have developed a technique for measuring the activity of different brain regions by measuring the amount of a radioactive analog of glucose that collects in them. Thin slices from the brains of animals injected with the radioactive analog are exposed to photographic paper. When the paper is subsequently developed, the darkness of an area will correspond to the intensity of radioactivity and thus to the concentration of the radioactive analog. In these pictures the darkness has been color-coded so that white represents the most intense activity, and red, yellow, green, and blue represent successively lower intensities. Because psychoactive drugs modify the activity of certain brain cells, this technique can be used to determine which regions are affected by the drug, information that is helpful in understanding both the nature of mental disorders and the effects of drugs on the brain. Figure 1 shows a cross section of the brain of a normal laboratory rat. The butterfly-shaped yellow and red structure in the center is the thalamus, a dense cluster of nerve cells important in relaying information to the cerebral cortex. Figure 2 shows the brain of a rat treated with the antipsychotic drug haloperidol. Note that the red regions in the thalamus have become less active. However, two white regions of intense activity appear near the top of the thalamus. It is likely that these are related to an undesirable side effect of haloperidol, movement disorders that resemble those of Parkinson's disease.

Strange Thoughts: The Disordered Mind

Senile plaques
Shown here are views through the microscope of the evolution of senile plaque (a small clot of protein and other chemicals) and neurofibrillary tangles (twisted bundles of abnormal protein fibers) in the brain during the course of Alzheimer's disease. The upper left photomicrograph (1) shows the first stage of plaque formation, a spherical zone outside the cells. No neurofibrillary tangles appear as yet. In the upper right photomicrograph (2) is an immature plaque surrounded by degenerating fragments of nerve cells. At a later stage (3) neurofibrillary tangles appear (arrow). A very late stage of evolution, shown at the lower right (4), is characterized by the absence of degenerating-neuron fragments.

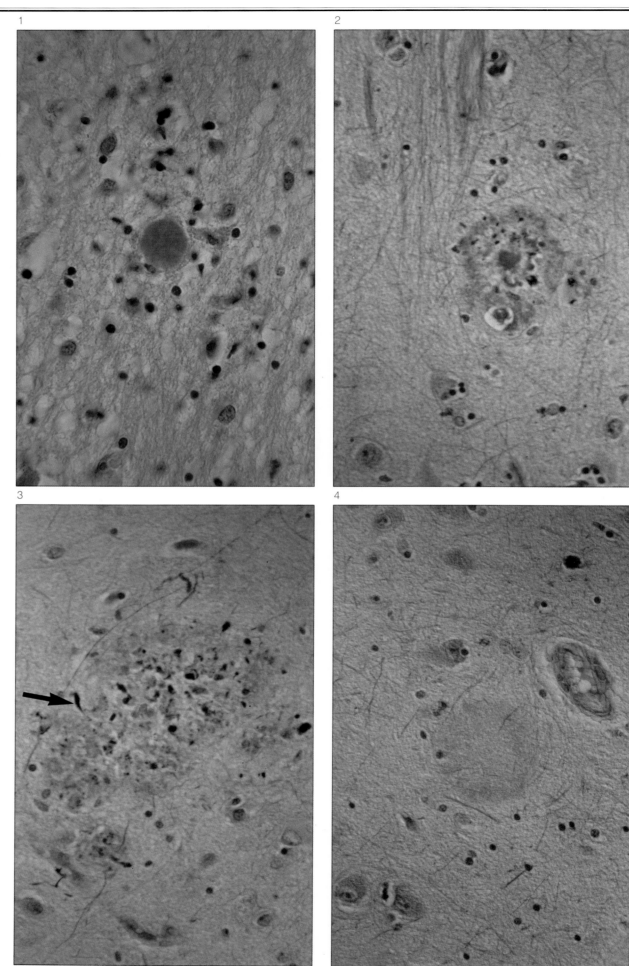

Strange Thoughts: The Disordered Mind

toward the end of the book and back toward the concepts of consciousness that began this part, as well as back to an earlier and almost incredibly naive time in the efforts of medical scientists to link the mind and the brain. The area I wish to look at in this transition is a dark area that bears the name psychosurgery.

The treatment bears the impressive name "transorbital lobotomy." It was used for curing depression, and in 1946 you could have it done right in the doctor's office. All he had to do was to stick an icepick over the top of your eyeball and twist it from side to side to separate the part of your brain known as the frontal lobes from the rest of it. If the doctor is conservative, he will do this to only half your brain at one time, waiting a week before plunging into the other side.

This procedure surely did something to relieve depression and psychoses, since some 5,000 similar operations were once performed per year in the United States. They also had dramatic effects on what might be politely called the "personality" of the patients treated in this fashion, freeing them not only from their prior mental problems but from much of their minds as well. *Life* magazine in 1947 described this surgery as "mental therapy." In the discussions to follow, I shall consider how horribly misguided this once-promising technique actually was, and I shall later consider the possibility of intervening somewhat more delicately into the puzzling relationship of the mind to the brain. Better answers will not come without better questions.

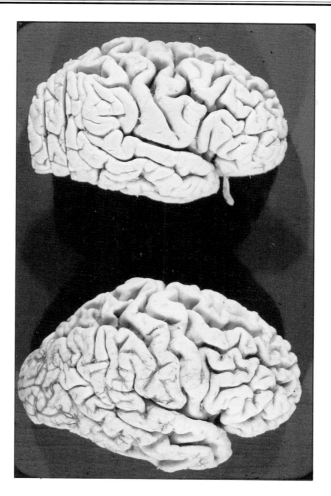

Familial Alzheimer's disease
Shown here are the brains of two individuals who had the familial form of Alzheimer's disease. The upper brain is that of the mother; the lower, her son's. Overall, the brains are shrunken, with a weight loss of about 9 percent, and the sulci (grooves) are widened. Atrophy is generally most pronounced in the frontal, parietal (upper rear), and temporal (bottom) lobes.

Strange Thoughts: The Disordered Mind

1

2

3

4

Strange Thoughts: The Disordered Mind

5

Yoga and its influence on the brain

Some researchers have found that changes in brain waves are associated with meditation. Does this reflect a changed state of the brain? How might this be determined? A few scientists have dedicated themselves to investigating these questions, and those who have provide only tentative answers. Among the meditation techniques best known in the West is Hatha Yoga, or Yoga of force. One of the many forms of Yoga, Hatha Yoga is based on a fanciful theory of physiology constructed by Hindu philosophers. It holds that a dormant divine force, kundalini ("serpent"), is situated at the base of the spine. A channel through the backbone, sushumna, leads through six psychic centers called cakras ("wheels") to the supreme psychic center at the top of the head, sahasrāra (a lotus with 1,000 petals). The aim of Hatha Yoga is to raise the feminine kundalini through the sushumna from one cakra to another until it unites with the masculine sahasrāra, attaining a state of transcendence. The union of kundalini with sahasrāra represents the union of the mind, freed of its bodily prison, with the universe around it (the original meaning of "Yoga" is union). To achieve this purpose, Hatha Yoga demands a rigorously disciplined development of will power: all automatic functions of the body must be brought under the control of the mind. Indeed, some yogis appear able to control their heartbeat, live for extended periods of time without breathing, and go without food or water for days. This remarkable control is achieved in part by assuming the positions illustrated here. Practitioners of Hatha Yoga have developed these and other positions to aid the meditation they claim is an integral part of mind control. The most famous position is the lotus (1). Whether or not it aids the flow of energy through the sushumna, the position at least helps the yogi to control her respiration. The forced spreading of the hips establishes a position of the pelvis and flexes the spinal column, both of which induce a feeling of relaxation. The yogi breathes slowly and regularly with a minimal displacement of her chest until her breathing is barely noticeable. Like all Yoga positions, the lotus minimizes the flow of sensory information to the brain. Increased blood levels of carbon dioxide may also play a role. Another position is called the back support (2). It is assumed while holding the breath after a normal inspiration. The yogi supports herself on the protruding surfaces of her heels with her knees extended. Her abdominal muscles are contracted and held, while her rib cage is expanded and held. This exercise is supposed to have stimulatory effects, helping to restore fast reflexes after a session of Yoga. The complete version of the inverted fish (3) is said to have an immediate effect on the secretions of the larynx and to enhance the circulation of the inner ear. The grasping of the big toe is based on the tenet of traditional Indian medicine that the second joint of the big toe has reflex effects on digestive blockages. Compressing the joint supposedly unblocks the small intestine. The simplified version of the inverted fish (5) lacks the big-toe effects. The side-pinch position (4) is assumed while holding the breath after expiration. The yogi flexes her torso to one side while contracting the muscles of her abdomen and grasps the big toe of her extended leg.

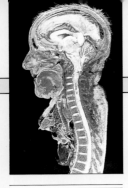

Understanding and Treating Anxiety

by W. Haefely,
J. G. Richards, and
H. Möhler, F. Hoffman-
La Roche and Co.

The 10 billion or so neurons of the human central nervous system (CNS) receive information from the outer world through the sense organs and from all parts of the body, via sensory neurons; command voluntary muscles via motor neurons; and regulate the activity of glands, involuntary smooth muscles (e.g., in the blood vessels and the intestines), and the heart via automatic neurons. Most of the central neurons, however, process incoming information and coordinate the commands given to peripheral tissues and organs. They also underlie such mental functions as thinking, learning, and remembering, and their activity determines moods and emotions (e.g., love, anger, anxiety).

The complexity of the human CNS results from the huge number of neurons, the enormous diversity of their morphology and chemical composition, and in particular, their extraordinary rich mutual interconnections. The thin endings of a neuron make tiny contacts called synapses with the cell body or processes of follower or target neurons. At these synapses the neuron "talks" to its target cell. The words of this language consist of chemical signal substances called neurotransmitters which are synthesized and stored in the nerve endings and released into the synaptic cleft upon a command from the cell body in the form of an electric impulse. On the surface of the follower cell the neurotransmitter is recognized and bound by receptor protein molecules that match the neurotransmitter as a lock matches a key. The receptor that has bound the neurotransmitter molecule transforms the message obtained by the signal substance into a change in membrane properties or in intracellular components. To highly oversimplify, the messages produced by the combination of the neurotransmitter with its receptor may be either positive (excitatory) or negative (inhibitory). The excitatory message tells the follower neuron to produce a command to its own target neurons; the inhibitory message tells the neuron to refrain from responding to excitatory input signals. The neuron continuously receives signals of partly opposite content; it continuously adds and subtracts the inputs; and if net excitation reaches a certain amount, an explosive response is triggered in the neuronal cell body. This response consists of a brief electric discharge of the membrane that travels into the nerve endings, where it initiates the release of a neurotransmitter. Theoretically, the CNS could operate with two neurotransmitters, one excitatory and one inhibitory. For yet unknown reasons, the CNS uses several positive and negative signal substances. In addition to these neurotransmitters, which use extremely quick-acting signal substances (0.001 to 0.01 second), neurons also release other active substances called neuromodulators. These neuromodulators usually do not by themselves mediate a positive or negative message but rather alter the responsiveness of their target neurons to excitatory or inhibitory neurotransmitters or both.

The most important inhibitory neurotransmitter in the mammalian CNS is the amino acid gamma-aminobutyric acid (GABA). About one third of all synapses in the brain are believed to be GABAergic, i.e., to use GABA as the signal. These GABAergic neurons represent the brain's own internal "braking system." This powerful inhibitory system is necessary because the CNS is so extremely sensitive to even smallest signals from the outer world and is so contin-

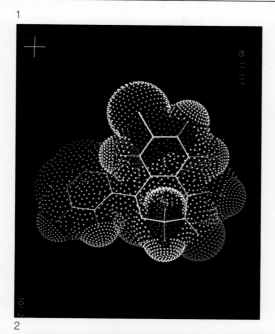

1. A structural analysis of diazepam
In this computer graphic, the dotted spheres of different colors represent the outer surfaces of different atoms.

2. The regional distribution of benzodiazepine receptors in a mouse brain (radiohistochemistry)
The receptors were made visible by intravenously injecting radioactively labeled benzodiazepine and exposing brain sections to a highly sensitive film. In this false-color computer image of a longitudinal section, the colors red, yellow, green, blue represent high to low densities, respectively, of benzodiazepine receptors.
(Magnification: x 2)

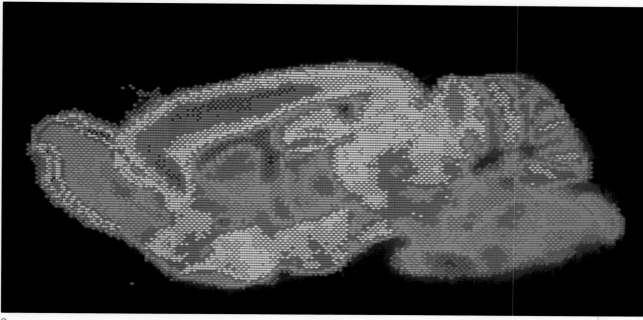

3. The cellular distribution of benzodiazepine receptors in bovine brain (immunohistochemistry)
The receptors were made visible by incubating brain sections with an antibody raised against the benzodiazepine receptor. In this light micrograph (interference-contrast image) of the dentate gyrus, four neurons and their processes are stained brown, which indicates the presence of benzodiazepine receptors on their surfaces.
(Magnification: x 100)

uously bombarded by information entering through the sense organs that we could not otherwise concentrate on any given activity or rest in the presence of the continuous flow of information reaching our brains. A moderate inhibition of GABA biosynthesis, release, or activity leads to restlessness, anxiety, muscle tension, convulsions, and eventually death. GABA inhibits neuronal activity by acting on a specific protein in the neuronal membrane, the GABA receptor. This GABA receptor is coupled to a transmembrane ion channel for negatively charged chloride ions. When GABA activates its receptor, the chloride channel is opened, and the normally very low conductance (or permeability) of the membrane for chloride ions is increased, which counteracts the effects of excitatory neurotransmitters. The GABA receptor and the chloride channel itself can be blocked by a number of agents that are convulsants because they reduce GABAergic inhibition and produce an uncontrolled hyperactivity of most CNS regions.

In 1956 the chemist Sternbach and the pharmacologist Randall, working in the research department of Hoffmann-LaRoche in Nutley, New Jersey, discovered the first active member of a large chemical family called benzodiazepines. This compound (chlordiazepoxide, trade name, Librium) and its most famous follow-up, diazepam (Valium), soon turned out to have a very broad and useful profile of activity. They reduce anxiety, tension, agitation, and nervousness; they attenuate or prevent convulsions and abnormal muscle tone; they facilitate sleep; and in higher doses they reduce attention and consciousness and prevent remembering experiences that occur in this state, making them useful adjuvants for smooth surgical interventions and the treatment of patients in emergency rooms. In spite of these numerous useful actions, the compounds have very low toxicity. The class of benzodiazepines was intensively exploited by the pharmaceutical industry, and at present around 50 benzodiazepine derivatives are available as tranquilizing, anticonvulsant, and sleep-inducing drugs.

In spite of intensive research efforts to understand the mechanism by which benzodiazepines act, it was only in 1975 that a group of researchers working at Hoffmann-La Roche in Basel, Switzerland, and one at the National Institutes of Health in Bethesda, Maryland, proposed on the basis of convincing experimental evidence that these drugs acted by enhancing the efficiency of GABAergic transmission. Two years later the presumed receptors for these drugs were identified by means of the highly specific binding of radioactively labeled benzodiazepines. Radioactively labeled benzodiazepines bound to their receptors in the CNS can be visualized in autoradiographic micrographs. These receptors are present throughout the CNS, however, at varying densities in different regions. Their distribution closely matches that of GABA receptors (identified by the binding of radioactively labeled GABA). This and the findings that benzodiazepines increase the binding of GABA to its receptors and that GABA enhances the binding of benzodiazepines to their receptors indicated that the GABA receptor and the benzodiazepine receptor are closely coupled physically and functionally. The isolation of the receptor molecules from neuronal membranes and binding studies with the isolated receptors made it clear that the GABA receptor, the benzodiazepine receptor, and the chloride channel form one molecular complex probably consisting of four subunits. When the GABA receptor is activated by GABA, a change occurs in the whole supramolecular complex: the chloride channel, which was closed, opens. When activating their receptors in the same complex, benzodiazepines increase the affinity of the GABA receptors for GABA, their natural ligand (binding molecule), and probably facilitate the channel-gating function of the GABA receptor. Hence, in the presence of a benzodiazepine, less GABA is required to open the chloride channel. Benzodiazepines thus enhance the efficiency of the GABAergic synapse; they act as a servo mechanism for the brain's own brakes. Since benzodiazepines do not increase the maximal inhibitory effect that GABA is able to produce, these drugs at the usual doses do not depress the brain's activity any more profoundly than the physiological GABAergic neurons do, e.g., in sleep.

Research in the benzodiazepine field was greatly stimulated by the previously mentioned discoveries, and in 1980 compounds were found that bind specifically to the benzodiazepine receptor but produce completely different effects than the classic benzodiazepine tranquilizers. One class of compounds has little or no effect when given alone, yet highly specifically blocks the effect of benzodiazepine tranquilizers. They are called benzodiazepine antagonists (or benzodiazepine receptor blockers) to distinguish them from the tranquilizing benzodiazepines, agonists of the benzodiazepine receptor. The other class of compounds produces effects that are in all respects the opposite of those seen with agonists, i.e., they induce anxiety, convulsions, and insomnia and increase muscle tone. They are called inverse agonists. While agonists facilitate the GABA-receptor function, and inverse agonists depress this function, antagonists do not affect the GABA-receptor function but block the access of agonists and inverse agonists to the benzodiazepine receptors.

Research on benzodiazepines has not only resulted in a detailed understanding of their mechanisms of action; it has also greatly advanced our knowledge of the role of GABA receptors in normal and abnormal states of the brain. The intriguing question of whether the brain produces its own benzodiazepines, i.e., a molecule that acts as an agonist or antagonist endogenous ligand of the benzodiazepine receptors to prevent or induce anxiety and other states, cannot yet be answered.

Benzodiazepine-receptor antagonists have proved to be highly useful in therapy to treat intoxications with benzodiazepine overdoses and in shortening the central depressant effect of benzodiazepines used in surgery or for diagnostic interventions. The knowledge of the molecular events at the complex of chloride channel, GABA receptor, and benzodiazepine receptor has led to the development of tailor-made ligands of the benzodiazepine receptor that combine agonistic and antagonistic properties. Because the density of benzodiazepine receptors varies on different neurons in the CNS that subserve various functions, there is good hope that such hybrid molecules (partial agonists) may retain full anxiolytic or anticonvulsant activity without producing the same degree of sedation (tiredness) and muscle relaxation as the classic agonists. Moreover, such partial agonists have shown little or no tendency to produce physical dependence in animal experiments, indicating that this may also be the case in humans.

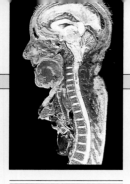

Senile Dementia and the Acetylcholine Connection

by J. M. Palacios, Preclinical Research, Sandoz, Basle, Switzerland

The discovery by Hornykiewicz and his colleagues of a clear neurochemical deficit, a marked decrease in the amine neurotransmitter dopamine in the brain of patients dying from Parkinson's disease, led to the first rational treatment of a brain disease. By administering L-dopa, the metabolic precursor of dopamine, to parkinsonians, it was possible to compensate for the missing transmitter and markedly improve the clinical state of these patients. The discovery ten years ago of a dramatic deficit of the neurotransmitter acetylcholine in the cortex and hippocampus of patients dying from Alzheimer's disease (AD) was followed by a number of clinical attempts to compensate for the missing acetylcholine with the hope that this would lead to an improvement in the symptoms of the demented patients. Many different approaches have been tried up to now. None has yet resulted in results to those obtained with L-dopa in Parkinson's disease.

An understanding of the rationale for the treatment of AD with cholinergic agents requires some information on the cholinergic synapse, the site of communication between neurons that use acetylcholine as a messenger. Few neurons in the brain use this chemical as transmitter. They send their axons to innervate different brain areas. A group of cholinergic neurons whose cell bodies are located in the lower (basal) part of the anterior brain (forebrain), more precisely in what is called the septum and nucleus basalis of Meynert, send their axons to innervate the whole cortex and the hippocampus. These neurons make acetylcholine from a precursor known as choline with the help of the enzyme choline acetyltransferase. Upon the arrival of nerve impulses, acetylcholine is released into the synapse and diffused to be recognized and bound to specific proteins in the postsynaptic neuron, the cholinergic receptors. The binding of acetylcholine to its receptors results in a cascade of biochemical and physiological effects that allow for the transmission of the synaptic message. The released acetylcholine is inactivated by an enzyme acetylcholinesterase, which degrades acetylcholine to yield choline once again. This choline can be taken up and recycled by the cholinergic neuron through a mechanism of uptake. For reasons unknown at the present, in the brains of AD patients the cholinergic neurons of the basal forebrain die which causes a deficit in communication through the cholinergic synapes. There are indications that this leads to a deficit in learning and memory. Agents such as scopolamine that will block cholinergic transmission by preventing acetylcholine from interacting with its receptors produce amnesia in experimental animals and healthy human volunteers. In trying to restore cholinergic transmission one can attempt to pharmacologically treat patients at three different levels of the synapse: (1) presynaptically in the remaining nerve terminals by stimulating the amount of acetylcholine the surviving neurons can form; (2) synaptically at the synaptic fosa by protecting the released acetylcholine from destruction by acetylcholinesterase, and (3) postsynaptically by acting directly on the postsynaptic receptors using artificial agonists (compounds that stimulate the receptor, as acetylcholine does). The three approaches have now been tested clinically. The approach presenting the fewest difficulties is the first one. Choline or its precursor lecithin have been given in very high amounts to AD patients without significant cognitive improvements being reported. Many reasons can underlie the failure of precursor therapy in AD. The main one is that, in contrast with the dopaminergic system, in the cholinergic neuron the availability of precursor is not a limiting factor for the synthesis of acetylcholine. Another reason could be that the number of surviving neurons is not sufficient to cope with the deficiency of transmitter.

Better or more promising results have been obtained with acetylcholinesterase inhibitors. In contrast to choline and lecithin these agents are more difficult to handle due to their toxic effects. Physostigmine has been given to AD patients in a number of controlled clinical trials and has been found repeatedly to improve AD symptoms. However, the use of this agent is limited by its short duration of action. New, more brain-specific, and longer-acting acetylcholinesterase inhibitors are currently being investigated.

Finally, in the third approach, directly acting postsynaptically muscarinic agonists have also been clinically investigated in AD. The basic requirement for the action of these drugs is the presence of its receptors in the brains of AD patients. Most of the investigations of the presence of muscarinic receptors in the brains of AD patients show that the disease does not seem to affect the number of postsynaptic cholinergic receptors.

Compounds such as oxotremorine, arecoline, and the spiro-piperidyl derivative RS 86 have been examined in AD. Oxotremorine is one of the most potent muscarinic agonists known. When given to humans, it produces many distressing side effects that preclude its clinical use. Arecoline, on the other hand, is too short-acting, although it has been shown to enhance learning in normal young volunteers and to ameliorate cognitive functions in AD patients. The most extensively tested muscarinic agonist is the compound RS 86. Until now clinical results with this drug have been mixed, with clear positive improvement in a minority of AD patients. As with the other agents, side effects such as salivation and sweating interfere with the clinical testing of this compound.

Clearly all the cholinergic agents that have been examined up to now present serious limitations for the definitive testing of what is called the cholinergic hypothesis of AD. New, more selective compounds, hopefully with fewer side effects, have to be developed before we know the usefulness of cholinergic substitution. Recent advances in the understanding of cholinergic mechanisms, such as the existence of different classes of cholinergic receptors in different organs and brain areas, are currently being examined with this goal in mind. The challenge of this research goes beyond the cure of AD, not an insignificant goal in itself. It touches the understanding of such basic scientific problems as the molecular basis of the highest brain functions: learning and memory.

1. The degeneration of nerve cells

The top photo shows the nucleus basalis of Meynert. The nucleus contains the cell bodies of the cholinergic neurons that innervate the neocortex. These neurons, seen in a healthy control in the center left photo, degenerate in Alzheimer's disease, as shown in the center right photo. This pathological lesion can be reproduced in laboratory animals by locally injecting a neurotoxin. In the right side of the bottom pair of photographs the cells of the rat nucleus basalis are visualized histochemically, while in the left side these cells were lesioned.

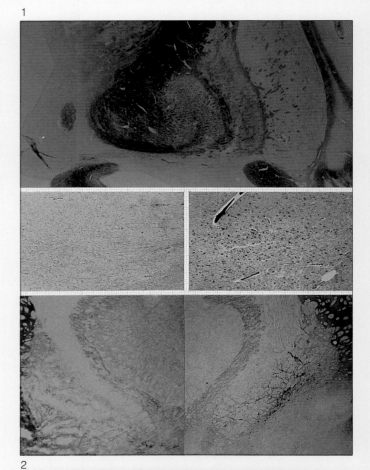

2. Muscarinic cholinergic receptors

False-color-coded autoradiograms illustrate the localization of muscarinic cholinergic receptors in the human brain. Areas with high densities of receptors are colored white or red, while low densities appear blue or violet. The top image shows the distribution of these receptors in the forebrain and illustrates the location of the nucleus basalis. The lower two autoradiograms show the cholinergic receptors in the human hippocampus. The left image is of a healthy control, while the right image is a severely demented patient. One can observe loss of cholinergic receptors in this patient.

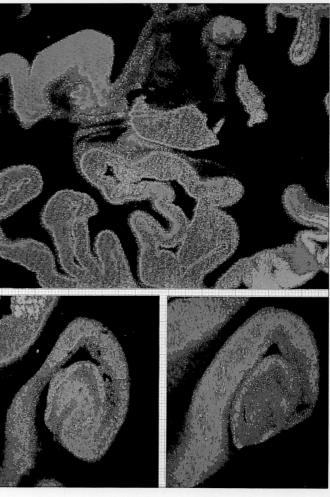

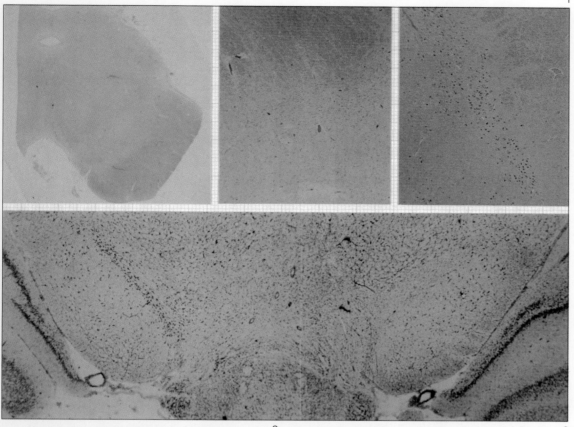

1. Parkinson's disease
In Parkinson's disease, cells in the substantia nigra degenerate. These cells use the chemical dopamine to communicate with other cells. The frame at the left shows a cross section of the human brain at the level of the substantia nigra. At the right is photomicrograph of the substantia nigra in a healthy brain. In the center is a corresponding photomicrograph of the brain of a patient with Parkinson's disease showing the loss of the cells containing dopamine. The lower figure shows the brain of a rat at the level of the substantia nigra. The left side is normal. The right side shows a brain injected with a toxin that destroys cells containing dopamine. The resulting condition mimics the effects of Parkinson's disease.

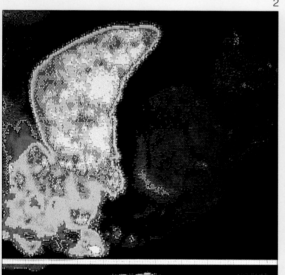
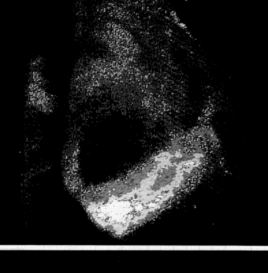
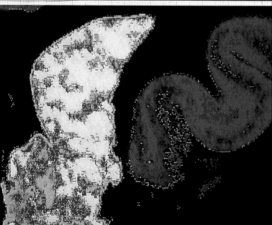
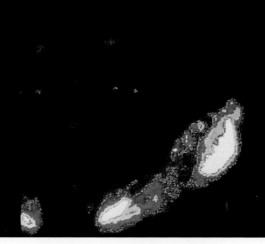

2. Dopamine receptors in the cat brain
Color-coded autoradiograms illustrate dopamine receptors in the cat brain. The top image shows D1 dopamine receptors, and the bottom image D2 dopamine receptors in the cat substantia nigra.

3. Dopamine receptors in the human brain
D1 (top) and D2 (bottom) dopamine receptors were visualized in the substantia nigra. High densities of these receptors were also seen in the caudate nucleus and putamen of man.

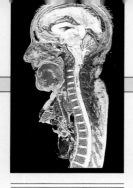

Toxins as Keys to Brain Communication

by James W. Patrick,
Baylor College of
Medicine, Houston, Texas

Information processing in the central nervous system requires extensive communication between neurons, and the proper functioning of the brain is critically dependent upon the fidelity of communication mechanisms. In general, animals are particularly sensitive to toxins that interfere with these mechanisms, and many plant and animal species have evolved toxins that either block or distort communication between neurons. The effects of plant toxins range from hallucinations and euphoria in the case of such toxins as LSD and cocaine to convulsions and death in the case of strychnine. A variety of snake neurotoxins produce paralysis and death. These toxins, which are obviously useful to the animals that produce them but pose a serious problem for man, are now being used by scientists to help understand how the brain works.

Toxins that block communication between neurons do so by interfering with specific molecular mechanisms. To those of us interested in how signals are passed between neurons in the brain, these toxins provide very powerful tools for the detection and analysis of this important process. We can use the toxins to reveal new communication mechanisms, and in many cases we can actually purify the molecules involved by means of their interaction with the toxins. An excellent example of this is research on the elapid neurotoxins, which produce paralysis by blocking the communication between motor neurons and muscles. In this specific case the communication is mediated by the release of a small molecule called acetylcholine by the nerve and its reception by an acetylcholine receptor on the muscle cell. The snake neurotoxin was found to bind tightly to the acetylcholine receptor. This prevents the interaction of acetylcholine with the receptor, which prevents excitation of the muscle and produces paralysis. This toxin provided a molecular tag for the acetylcholine receptor and assisted greatly in its purification and subsequent study.

The acetylcholine receptor was the first neurotransmitter receptor to be purified, in part because the toxin provided such a powerful tool and in part because there was in nature a rich source of the receptor itself. Various electric eels and rays generate their electricity by the simultaneous activation of a very large number of acetylcholine receptors present in a specialized electric organ. The fortunate conjunction of the snake neurotoxin and the electric organ permitted the actual purification of a neurotransmitter receptor. An initial problem in the purification of the receptor was knowing whether the molecule purified was in fact the receptor molecule. Since the function of the receptor in animals is to bind acetylcholine and initiate an electrical event in the muscle cell, it was necessary to devise a similar sort of process in the laboratory. After many years of work, the purified receptor was finally inserted into an artifical membrane, where its function could be measured. Once it was known that the receptor had been purified, it was possible to begin studies on its structure and function.

The acetylcholine receptor was found to be composed of four different protein molecules called alpha, beta, gamma and delta. These area assembled in groups of five composed of two alpha subunits and one each of the other subunits. The neurotransmitter acetylcholine activates the receptor by binding to the two alpha subunits. The receptor was found to pass through the cell membrane in an arrangement that allows the interaction with acetylcholine to create a channel or pore in the membrane. The passage of electrically charged particles through this channel initiates the electrical events in the muscle cell. The general properties of the receptor were becoming clear, but its detailed structure and mechanism of action remained obscure.

The purification and reconstitution of the acetylcholine receptor from the electric organ made possible subsequent studies that focused not on the protein but on the genetic material that specifies the structure of the protein. This new approach to understanding communication in the nervous system has changed the way we now think about the study of the brain. In this approach, recombinant DNA technology is used to isolate the genetic information that determines the structure of the receptor subunits. Once isolated, this information can be duplicated and made available for a wide variety of studies. In the case of the acetylcholine receptor the complete amino acid sequences of the four proteins that compose the receptor were deduced from recombinant DNA clones. The knowledge of these sequences allowed the creation of specific models of the receptor that made specific predictions about how altering certain parts of the molecule might affect receptor function.

The isolation of the recombinant DNA molecules encoding the acetylcholine-receptor subunits also made possible a means of testing theories about the receptor. From these recombinant DNA molecules we can transcribe messenger RNA molecules that directly instruct protein production. This RNA can be injected into frog egg cells, where it will direct the synthesis of a functional acetylcholine receptor molecule. We can then assay the function of the receptor using simple physiological techniques or, once again, its interaction with the snake neurotoxins. The advantage of these techniques comes not from simply being able to produce a receptor from the cloned genetic information but from the fact that we can now alter the genetic information at will and thus create different, novel acetylcholine-receptor molecules. By choosing which parts of the molecule to alter, we can produce receptors that have altered functions. This allows us at last to study the relationship between the structure of the molecule and its function as the receiver of signals from nerve cells.

A second major advantage resulted from the isolation of recombinant DNA molecules encoding the electric organ acetylcholine receptor. We could then use these molecules to isolate other genes encoding other, rarer receptor molecules. In the beginning this technique was used simply to isolate clones coding for acetylcholine receptors found at neuromuscular junctions in mice, calves, and humans. More recently, however, it has been possible to isolate recombinant DNA clones encoding neurotransmitter receptors that subserve the communication between neurons in the brain. Study of these clones has revealed that the brain contains an entire family of neurotransmitter receptors related to the one found at the nuromuscular junction. All seem to be descendants of some more primitive molecule that has apparently evolved to play many roles in the nervous system This discovery may be of tremendous value in our efforts to understand how neurons in the brain communicate with one another.

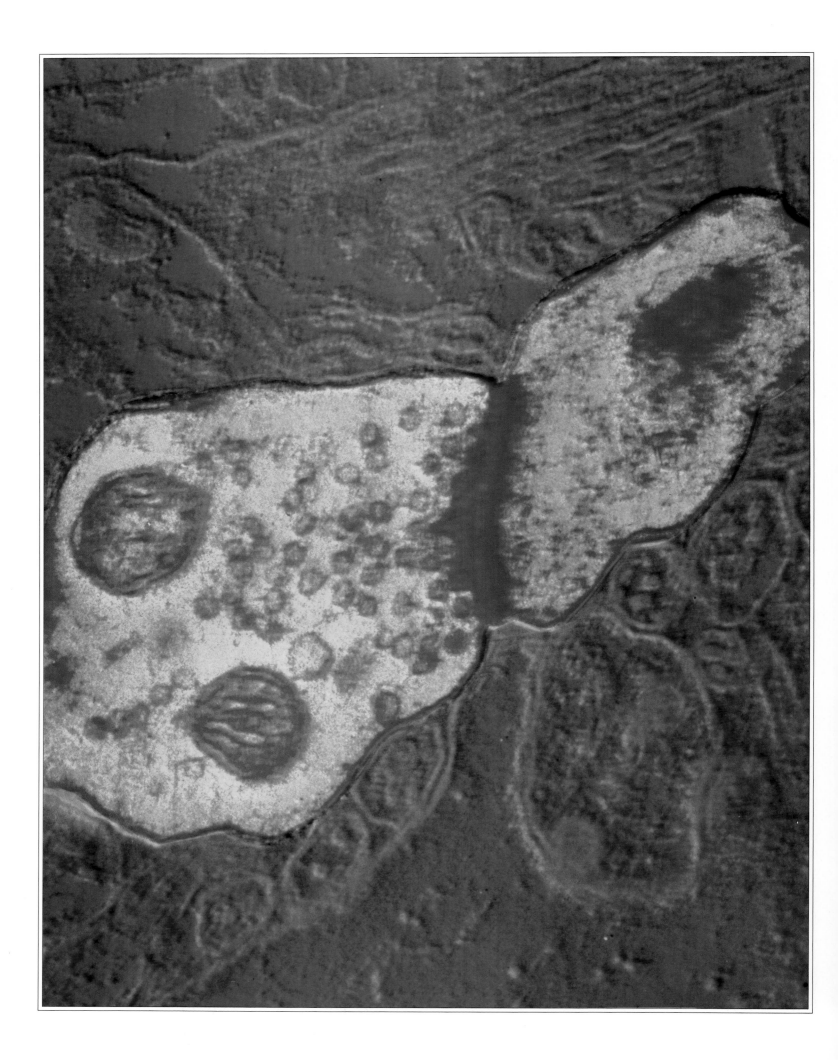

8
Paths through the Brain: Maps for Explorers

The landscape of the brain is not really very different from that of the earth, the moon, or any other object that can be represented on a piece of paper or a computer screen. It has hills called gyri and valleys called sulci and many other landmarks that can be used to guide navigation. Most have Latin names that can drive students of neuroanatomy to tears. Brain mapping was a scientific rage in the nineteenth century. Explorers vied avidly with one another to have a region named after them. The rage reached its peak with the phrenology of Francis Joseph Gall. This "modest" title for one of his works, originally written in French, tells it all: *The Anatomy and Physiology of the Nervous System in General, and the Brain in Particular; with Observations on the Possibility of Determining Many Intellectual and Moral Dispositions of Men and Animals by the Configurations of Their Heads.*

When a more scientific attitude came to prevail, regions of the brain got numbers. A scheme developed by the German neurologist Korbinian Brodmann in 1909 remains in favor today, at least as far as identifying to scientific colleagues where you have staked out your territory. "I'm working on area 17a" will tell your colleagues what sort of person you are just as much as if you said "I'm from Brooklyn" or "I was born in Los Angeles." A map of Brodmann's areas looks like a paint-by-numbers picture, with patches of brain surface identified by numbers, each patch representing an area that is more or less distinct in terms of its microscopic structure, and sometimes its function.

You can hold a human brain in your hand and examine its features, much as you might look at a satellite map of the earth and identify its continents. You can look more closely at the "continents" of the brain and identify them with some of their principal functions: vision belongs to the territory toward the back, hearing involves the parts of the brain just below the great valleys along the sides. You can take sections of the brain and put them under a microscope and see treelike nerve cells, surrounded by glia, whose supporting functions are not well understood. And you can make maps of how cells are connected together like components in a TV set or a computer. These maps, as we shall see, are very handy for finding your way around and are not unlike a street map of a city.

Despite the formidable Latin names and the microscope slides that look like forests, neuroanatomy is no more formidable than a visit to a strange and rather exotic city. Let's start with a "satellite view." It shows a pair of cerebral hemispheres, the cerebellum, the brain stem, the spinal cord, and the peripheral nervous system. The brain and spinal cord constitute a central information processing system. It receives information from the world through the peripheral nervous system. Sensory nerves send in messages from sense organs on the surface of the body. Motor nerves carry commands from the brain to muscles.

Closer examination of the brain reveals some of its principal operational units. The cerebral cortex seems to be the highest-level information-processing center and reaches its most extensive level of development in the human. It is involved in thinking, planning, perceiving the details of sensory information, and issuing commands to muscles. All the senses except smell send information to the cortex through the thalamus at the center of the brain. Commands to muscles go through large bundles of motor fibers to the spinal cord, but the cerebellum seems to be essential in keeping these commands in order. Another subcortical system with cell clusters, the basal ganglia, is also involved in motor control. Its effects are most noticeable when they go wrong in Parkinson's disease.

Other important parts of the brain are the hippocampus, which is critical in memory; the limbic system, which mediates emotions; and the hypothalamus, which controls essential life-support activities like eating, drinking,

The synapse
This artificially colored photomicrograph shows a vital structure of the brain, a synapse, or a connection between two neurons. An electrical signal arrives on the nerve fiber at the left, triggering the release of a chemical neurotransmitter stored in small packets (vesicles), which appear as small red circles, into the gap separating the fiber from the nerve cell at the right. When the neurotransmitter attaches to a receptor on the second cell, it may increase or decrease the level of activity in the receiving cells, according to the nature of the synapse.

Paths through the Brain: Maps for Explorers

The sinus of Breschet
Gilbert Breschet (1783-1845) was particularly interested in the system of veins in the bones of the head when he was appointed chief of anatomy at the Hôtel Dieu (the oldest hospital in Paris). He devoted himself to preparing an atlas in color to be produced by the then newly developed process of lithography. This plate from his atlas shows what was called for almost a century the sinus of Breschet, a system of veins that collects blood from the brain. The study of the sinus of Breschet has recently been taken up again by Roger Saban in the course of his paleontological research. With the aid of casts made from fossilized human skulls, he has shown that the sinus originated in a simple form in Australopithecus. *In the course of evolution it became more complex until it reached the vascular partitioning that covers the entire parietal surface in modern humans. Only in Neanderthal man is the large anterior vein (the sinus of Breschet) always present. In modern humans, ancient forms of the system of veins sometimes recur. (Notes by Roger Saban, professor at the Museum d'Histoire Naturelle, Laboratoire d'Anatomie Comparée).*

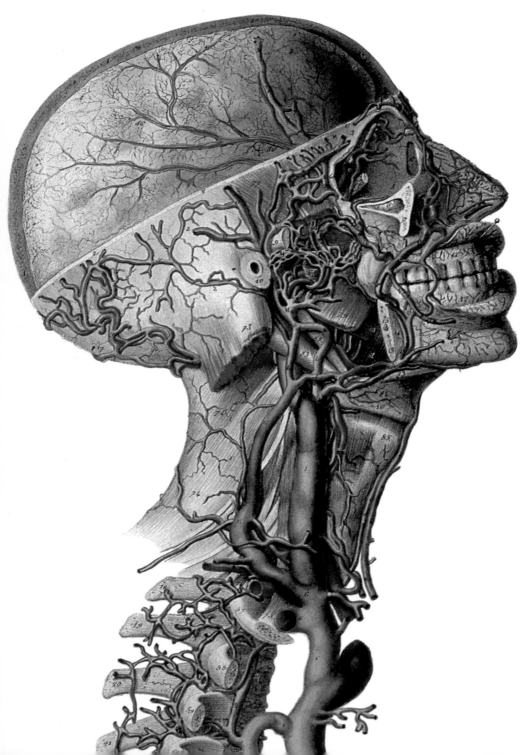

temperature regulation, and reproductive activity. There are many more detailed features of the brain that seem to be very important, but they are not at all well understood.

The basic elements of the nervous system are nerve cells, or neurons. Each cell is a self-contained, living chemical factory and information-processing system. Typically, a neuron has a receiving end and a transmitting end. At the receiving end it has a treelike system of dendrites, which receive information from other neurons. When a neuron is activated by other neurons, it sends a brief electrical pulse lasting about a thousandth of a second along its output fiber, the axon, which in turn releases neurotransmitters that affect other neurons. A neuron in the brain may receive signals from thousands of other neurons and may in turn send its messages on to thousands of others. Since there are about a trillion neurons in the brain, the number of connections among them is staggering.

Walle J. H. Nauta of the Massachusetts Institute of Technology has been one of the leaders in developing methods that for the first time make it possible to trace the connections among specific cells in the brain through the otherwise impenetrable maze of tangled nerve cell connections. Early neuroanatomists were able to apply to thin slices of brains chemicals that react with various substances present in neurons to stain them and thus make them visible under a microscope. But these stains colored only the bodies of cells. They did make it possible to see that there are clusters of neurons, each known as a nucleus, scattered throughout the brain. They could not, however, reveal the nerve fibers thought to carry messages between nuclei.

The problem is a fascinating one, not unlike trying to figure out how a computer works by tracing all the connections among its components. Will the wiring diagram of the computer help us to figure out how to program it? Will the wiring diagram of the brain help us to learn what really constitutes consciousness? That is a question that we are not yet wise or clever enough to answer. As will be pointed out in the discussion of how the visual system works, knowing the wiring diagram of that system is proving essential to understanding how it works. What was once considered a dry, dull, and hopelessly confusing subject for study, the anatomy of the brain, is today proving not only an essential but also a dynamic and exciting part of the story of how the brain lives.

The techniques that Nauta and others have developed make it possible to trace axons in either direction, either from the cell bodies to the contacts with other cells or from the terminations back to the cell bodies. This means that a stain can be injected into a nucleus at one point in the brain, say the medial geniculate nucleus, which receives signals from the retina, and the axons of these fibers can be traced to the visual cortex, where they make connection with other cells. This is known as anterograde tracing. If we wish to work in the retrograde direction, a stain can be applied to an area of visual cortex where cells seem to respond particularly to motion. The special stain will then travel back along the axons to the cell bodies and identify the source of the motion information in the primary visual cortex.

The very newest "stains" are not stains at all, but labels for specific chemicals present in nerve cells provided by the very powerful tools of molecular biology. One technique, with the daunting name of immunocytochemistry, takes advantage of substances known as monoclonal antibodies, which can act as bloodhounds to home in on a target molecule that is hidden in a forest of many millions of other chemicals. Application of a monoclonal antibody to brain slices makes it possible to locate exactly where in the brain there are cells that contain a particular transmitter, receptor, enzyme, or other target molecule. In view of all the great diversity of potential transmitters that are now being discovered in the brain (see chapter 12), immunocytochemistry is becoming an essential tool for finding one's way around in the brain.

While enormously useful, all of the techniques discussed above yield results only after the brain is no longer living. Furthermore, they provide only static pictures of events and processes that took place in the past. They reveal little about the dynamics of thinking, listening, or speaking. Exciting new ways of peering into the functioning human brain are rapidly changing this situation and providing neuroscientists with increasingly fascinating insights into the brain's machinery.

Standard X-rays have long been used to look inside the living human head, but the resulting pictures were crude and revealed only gross structural features or abnormalities.

The first major improvement came in the early 1970s with the development of computer aided tomography (CAT). A narrow beam of X-rays is aimed through the head at a detector on the opposite side. The source of the beam and the detector move about the head so that the beam makes an imaginary slice through the brain. Different structures in the brain block the beam to different degrees and a computer stores the detected intensity for each path. When the scan is complete, the computer calculates the relative density of the brain at each point and displays a picture that looks just like a slice through the brain.

CAT scans provide very sharp images of the brain and distinguish well between white and gray matter. They can be used to detect brain atrophy, the accumulation of fluid, the loss of myelin (the electrical insulation surrounding neurons, which deteriorates in multiple sclerosis), and many other abnormal changes in brain tissue. They are thus highly valuable clinical tools.

The first technique to reveal chemical activity inside the brains of awake human beings is a new development called positron-emission tomography (PET). It is based on radioactive decay that releases subatomic particles called positrons, the antimatter counterpart of electrons. A positron almost immediately collides with an electron, converting the mass of the two particles into two pulses of energy that travel in opposite directions. Detectors arranged in a ring around the head record a decay when two detectors directly opposite one another receive an energy burst at almost the same time. A computer calculates the location of the decay and prints out a map of the concentrations of radiation sources.

Many PET studies use a radioactive sugar that is taken up by active brain cells. PET scans made with this sugar show that the regions of the brain involved in hearing light up when a person is listening to a story. If the listener is also asked to remember the story, brain regions involved in memory storage light up as well. PET studies have also shown that brain regions involved in epilepsy consume excess energy, as do some tumors. In addition, PET promises to help us diagnose Alzheimer's disease and understand brain abnormalities involved in schizophrenia and other psychiatric disorders.

Paths through the Brain: Maps for Explorers

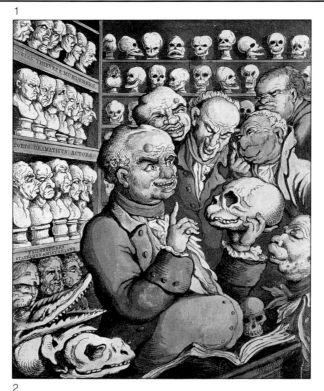

1. Phrenology
An English cartoon from the early nineteenth century illustrates the practice of phrenology, the study of the attributes of one's mind according to the shape of one's skull. Phrenology was devised by the German physician Franz Joseph Gall.

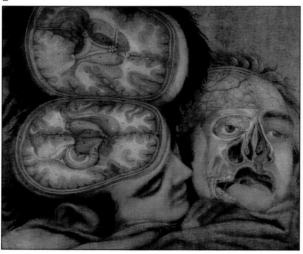

2. Dissection of the head
An illustration from J.-F. Gautier d'Agoty's Anatomy of the Head in prints, *published in French in 1748.*

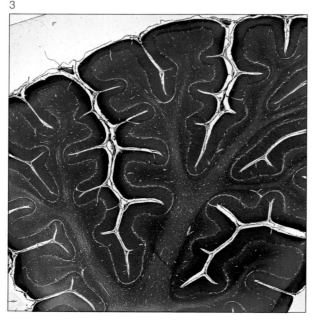

3. The cerebellum
This section of the cerebellum is stained using the Cajal technique, developed early in this century by the pioneering neuroanatomist whose name the technique bears.

An altogether different way of looking into the living brain is based on the ability of certain atoms in the brain to act like tiny magnets. If the head is placed in a very strong magnetic field, these atoms line up parallel to the field and spin at a rate corresponding to a radio frequency. Application of a radio signal at this same frequency adds energy to the spinning atoms. When the applied signal is switched off, the spinning atoms return the absorbed energy in a brief radio pulse. Manipulating the strength of the applied magnetic field allows a computer to reconstruct an image that reflects the distribution of the magnetic atoms. Since different brain parts have different concentrations of these atoms, the image reflects the brain's structure and does so in finer detail than any other imaging technique.

This technique, known as magnetic resonance imaging (MRI), has been used for studying human brains for only a few years but has already produced some interesting results. It has confirmed the finding that the brain region associated with language in the left hemisphere is larger than the corresponding area in the right hemisphere and has produced evidence against sex-related differences in the corpus callosum. It has also revealed changes in brain structure associated with several psychiatric disorders. The functional implications of these structural abnormalities pose a challenge to future research.

A variation of MRI, termed magnetic resonance spectroscopy, holds great promise in its ability to measure the amount of a specific chemical present in any region of the brain. The resolution of MRI is sharper than that of PET, so rapid advances in the field should soon make it possible to uncover biochemical changes associated with both normal brain activity and the abnormal activity underlying psychiatric disorders. Dynamic chemical brainscapes of the living human brain thus promise to be the ultimate maps for explorers of the brain.

Paths through the Brain: Maps for Explorers

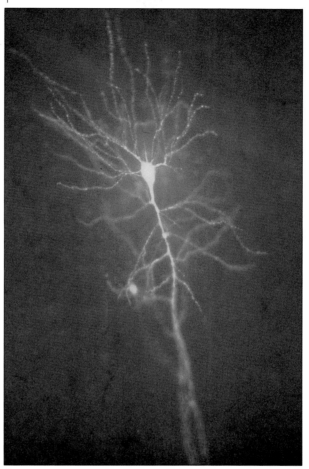

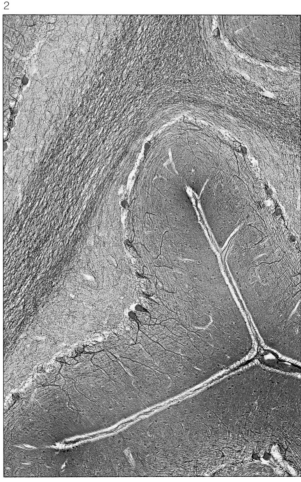

Neurons
Above (1), a nerve cell injected with a fluorescent dye shows its transmitting fiber, the axon, extending toward the bottom and its receiving structures, dendrites, branching above and to the sides. In the lower plate (2), axons stream in from above, sending branches to an orderly row of neurons arrayed in an arc.

Paths through the Brain: Maps for Explorers

1. Myelencephalon (medulla oblongata)
2. Metencephalon (pons and cerebellum)
3. Mesencephalon (midbrain)
4. Diencephalon (thalamus)
5. Telencephalon (cerebral hemisphere)
6. Cortex of the temporal lobe (T)
7. Third ventricle
8. Lateral ventricle
9. Anterior columns of the trigone
10. Corpus callosum
11. Interhemispheric fissure
12. Sylvian sulcus
13. Lateral ventricle

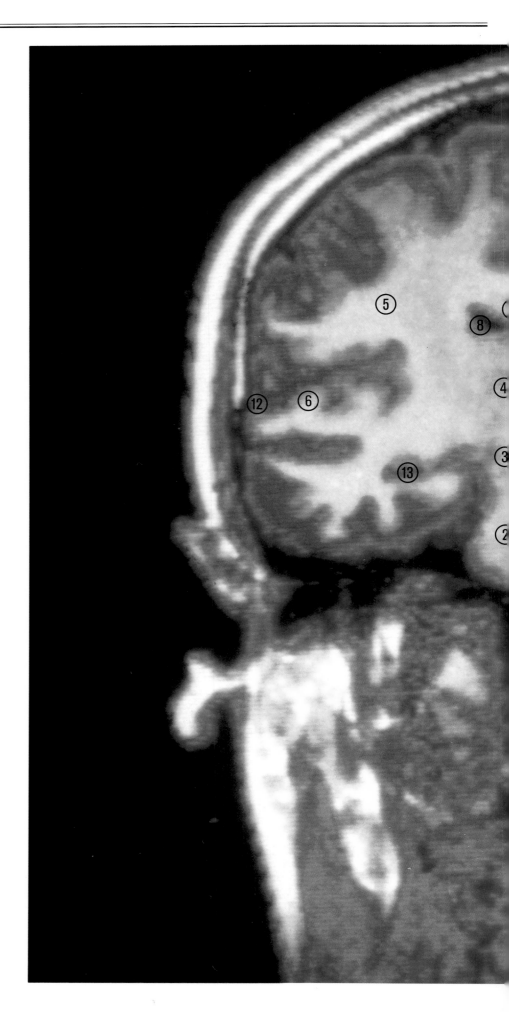

Paths through the Brain: Maps for Explorers

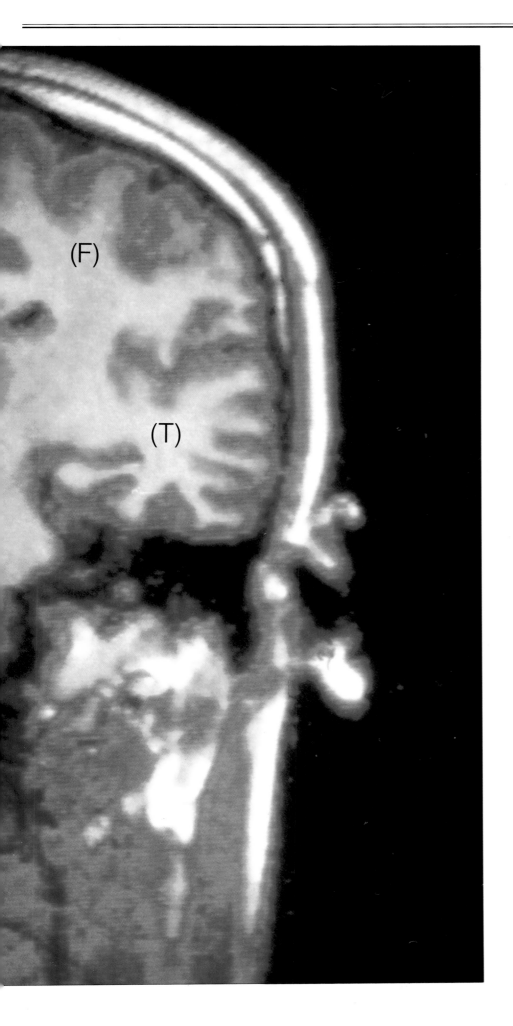

113

Paths through the Brain: Maps for Explorers

1. Cervical spinal cord

2. Myelencephalon (medulla oblongata)

3. Metencephalon (pons and cerebellum)

4. Mesencephalon (midbrain)

5. Cerebellum

6. Fourth ventricle

7. Sylvian aqueduct

8. Third ventricle and diencephalon (thalamus)

9. Hypothalamus

10. Hypophysis (pituitary gland)

11. Chiasm of the optic nerves

12. Corpus callosum

13. Trigone

14. Occipital cortex

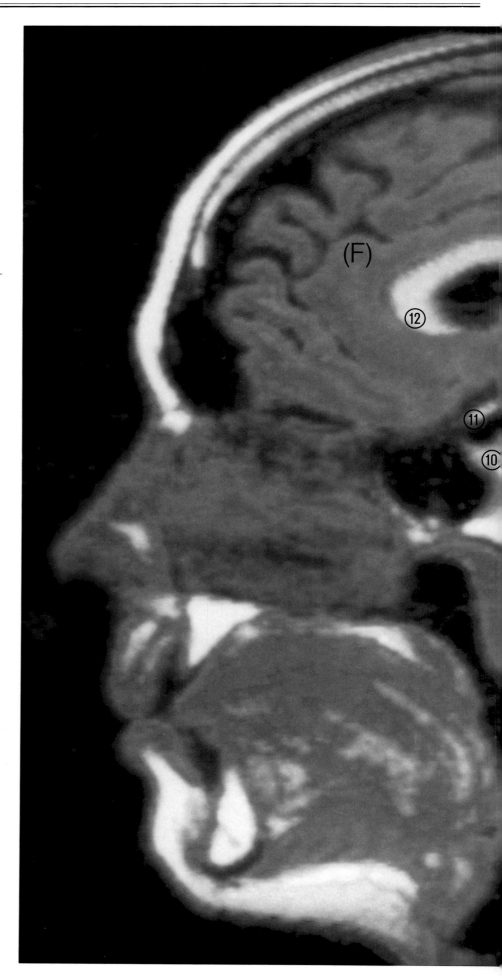

Paths through the Brain: Maps for Explorers

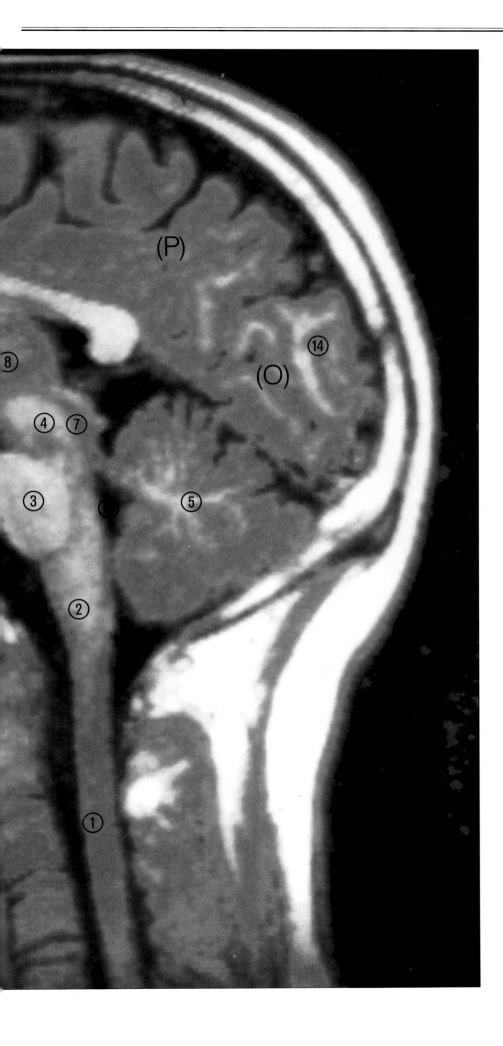

The human brain
Frontal (pp. 112-113) and midsaggital sections of the living human brain made with magnetic-resonance imaging and color-coded by computer. The motor cortex is situated just anterior to the central sulcus, the somesthetic sensory area just posterior to the central sulcus, the visual cortex in the occipital region, and the auditory cortex is buried in the floor of the Sylvian sulcus. The letters F, P, O, and T designate the frontal, parietal, occipital, and temporal cortex, respectively.

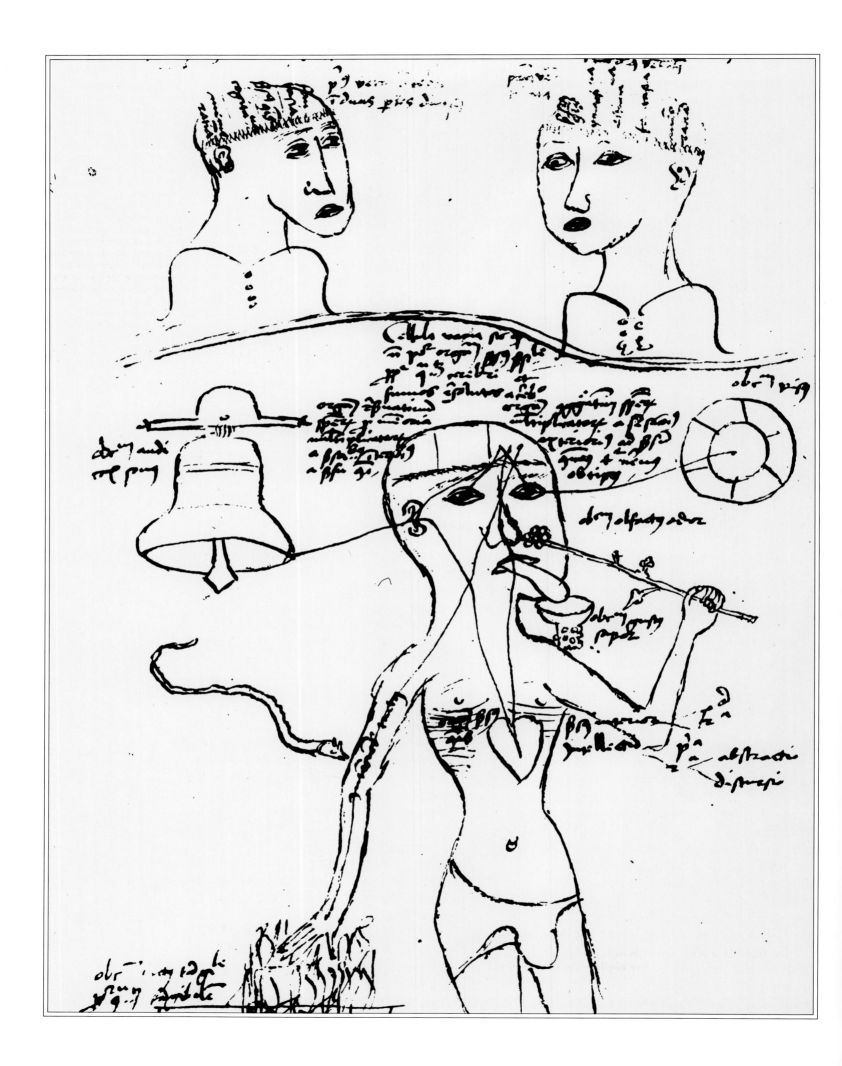

9
Is It Real?
What Our Senses Tell Us

Human beings have evolved as organisms that are principally dependent for survival on their brains rather than on physical strength or speed. We therefore occupy a peculiar position on earth. If we accept the notion that our conscious self is dependent on the brain for acquiring and interpreting information about what is going on in the world around us, how reliable is our source? This is, of course, just another way of posing the old philosophical puzzle of how we know what reality is.

Brain research is giving new life to this question, possibly even raising it to very high levels of importance in determining the future of the world in the face of problems of international understanding. Brains are extraordinarily complex biological systems that modify themselves as they grow (see chapter 11 for a discussion of how this occurs). A very important factor in determining how the brain develops is the experience it goes through as it is growing. The brain's view of the world is through a few rather limited physical stimuli that affect the skin, the eyes, the ears, the nose, and the tongue. Our bodies are immersed in a sea of radio, television, and other electromagnetic waves that we cannot see or feel. We cannot see the infrared or the ultraviolet radiation of the sun, nor can we hear the ultrasonic cries of bats. Our senses are indeed limited.

All sensory systems serve a similar purpose, to bring information from the outside world into the brain. The general patterns of their organization are also similar, but there are some important differences. In each case a sense organ receives a signal: sound in the ear, light in the eye, and so on. It translates this signal into the language of the nervous system and sends the resulting message coursing toward the brain. Different parts of the brain are specialized for interpreting the messages from different sense organs. We know many of the details about how some of these systems work, particularly for seeing and hearing. Others, like the sense of smell and the sense of pain, are more elusive.

One of the most studied senses is vision. Scientists have carefully unraveled the connections of brain cells in the visual system and have studied how they respond to light, so we have many clues about how the brain takes visual images apart. What is particularly elusive, however, is how the brain puts the pieces back together, turning two-dimensional patterns of light on the retinas into our perception of the visual world. In one case, however, the perception of color, we are beginning to get a good idea of what the brain is doing.

Most people think that the balance of red, green, and blue light reflected from an object into the eye determines its color. It is easy to demonstrate that this notion is not true, however, simply by noting that objects remain the same color in daylight, fluorescent light, and incandescent light, each of which contains a mix of wavelengths of light very different from the others. Edwin Land, inventor of the instant camera, has provided an explanation of this phenomenon in what he calls the retinex theory, a term that combines "retina" and "cortex" to suggest that both parts of the visual system are involved in perceiving color.

Retinex theory proposes that the retina and the cortex cooperate to perform some complex computations on the basis of light received from all areas within the visual landscape. A separate computation is carried out for each of three wavelengths of light that correspond to what we normally think of as red, green, and blue, the wavelengths to which the three types of receptors in the retina are most sensitive. According to the theory, the color that we perceive at a particular location is determined by three numbers, computed by dividing the amount of light received from that location at each wavelength by a weighted average of the amount of light at that wavelength received from all parts of the field of vision. The weighted average gives more weight to light coming from close to the location in question than to that coming from far away. The three numbers are coordinates in a

Epitome
An ink drawing from Epitomata, *by G. de Hardewyck, presenting the functions of the senses. It was published in Cologne in 1496.*

Is It Real? What Our Senses Tell Us

A bat
Bats, like other animals, have sensory capabilities not found in humans, such as the ability to hear ultrasonic frequencies.

color space of three dimensions and uniquely determine the color we see, just as the three dimensions of physical space uniquely define the location of an object. Land has conducted a number of experiments showing that the numbers computed in this way correctly predict what color an observer will see under a number of unusual lighting conditions.

This remarkable theory suggests that our visual systems evolved so that we see the colors of objects as the same, regardless of the mix of wavelengths of light falling on our retinas. Furthermore, this complex computation is carried out virtually instantaneously without our even being aware of it. There is also evidence that retinex computations emerged quite early in evolutionary history. David Ingle of Northeastern University has shown, for example, that goldfish brains apparently perform retinex calculations, allowing the animal to select a specific color despite changes in the mix of wavelengths of light.

Other experiments have found evidence within the brain that it does indeed perform the required retinex calculations. S. Zelki of University College, London, has identified cells in the cortex of monkeys that fail to respond when light of a certain wavelength falls on the retina, yet this same wavelength must be present for the cell to respond to other wavelengths. Apparently, these cells are somehow involved in calculating locations in color space. In other experiments David Hubel and Margaret Livingston at Harvard University have found cells in cat cortex whose activities also suggest involvement in retinex computations. The evidence in support of the retinex theory is therefore quite strong and leads us to the conclusion that our perception of color is not affected by the amount of light received, by anything we have learned, by the surroundings, by the adaptation of the eye, or by anything else that does not affect the location of the point in color space determined by retinex computations.

While we have few insights into how cells in the human brain interpret visual information, research on the psychology of seeing is providing some very good clues about what to look for. Vilayanur S. Ramachandran at the University of California, San Diego, has developed what he calls the utilitarian theory of perception. It holds that the human visual system has acquired a bag of tricks over the course of millions of years of evolution and that these provide it with different kinds of short cuts to pick out useful features of the visual world. For animals in the wild, catching food and avoiding being eaten are essential for survival, and anything that moves is a candidate for being prey or predator. It therefore makes sense that some of the tricks that proved useful for survival might be associated with the rapid detection of motion.

Ramachandran looks for short cuts used by the visual system in studies of apparent motion. When we watch a movie, we are experiencing apparent motion. Each image flashed on the screen is a motionless scene, but one follows another so fast with such a small change in the position of each image that our visual system tells us the images are moving. By manipulating images that appear to move, Ramachandran believes he has found some clues to one of the tricks of the visual system. He argues that such perception occurs so rapidly that there is not time for the brain to perform any elaborate comparison between successive images. Therefore, at some early stage of processing visual images, there must be some system that rapidly identifies some correspondence between successive images and associates that correspondence with something that moved. We do not have to think about it. Any creature who spent too much time identifying two successive glimpses of a shape moving through the trees as being a moving leopard would not survive for long.

One of the simplest ways to demonstrate apparent motion is to flash two spots of light one after the other fairly close together on a screen. It will appear to be one spot moving back and forth. Using variations of this and other simple displays, Ramachandran deduced several simple laws about the perception of apparent motion. These include the fact that the visual system most quickly detects large bright or dark blotches or textured patches before it pays attention to finer details like lines or edges. The visual system seems to have its own laws of physics, preferring to interpret ambiguous movement as conforming to the most physically likely pos-

Vision
The special glasses shown here deliver computer-controlled visual images to the eyes. Positron emission tomography is then used to measure the activity of brain regions stimulated by the images.

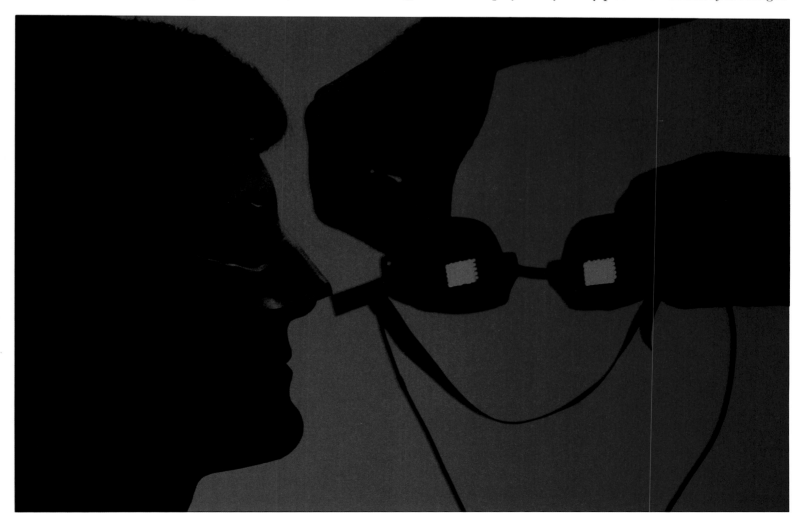

sibility. For example, if a pattern of flashing spots can be interpreted as two objects moving in parallel or two objects passing through each other, we see it as two objects moving in parallel rather than through each other. If, however, the spots are shown in three dimensions so that the paths that cross are at different depths, the crossing that is now physically possible does appear. Finally, unless there are contrary indications, all objects in an area appear to move together, even if some are not moving, a phenomenon known as captured motion.

There has so far been little research done to find out how these recently identified processes occur in the brain. Recent research on the visual system does indicate, however, that there are separate cells that respond to the motion of large blobs, while others are sensitive to color and to finer details like angles and the ends of lines. Since the phenomena of apparent motion occur before the images have a chance to reach our conscious awareness, research on these phenomena is not likely to benefit, or suffer, from our intuitions about how we perceive apparent motion.

Information from our nose gets into our brain in a way quite different from the information from other sense organs. It has privileged access in a sense, since neurons in the nose go directly to the cortex. Everyone, of course, is aware that the sense of smell is quite a primitive one and was developed quite early in the course of evolution. So in a sense, the study of smell takes us back to a time when smell was more important to survival than it is to us humans today. Smell takes up most of the brains of fish. But only a relatively small portion of the human brain, buried underneath and toward its front, is devoted to information from this sense. For mammals, especially pigs, the sense of smell is vital for mating. Some

Touch
Gabrielle d'Estrées, mistress of Henry IV of France, and one of her sisters were immortalized by a painter of the school of Fontainebleau at the end of the sixteenth century.

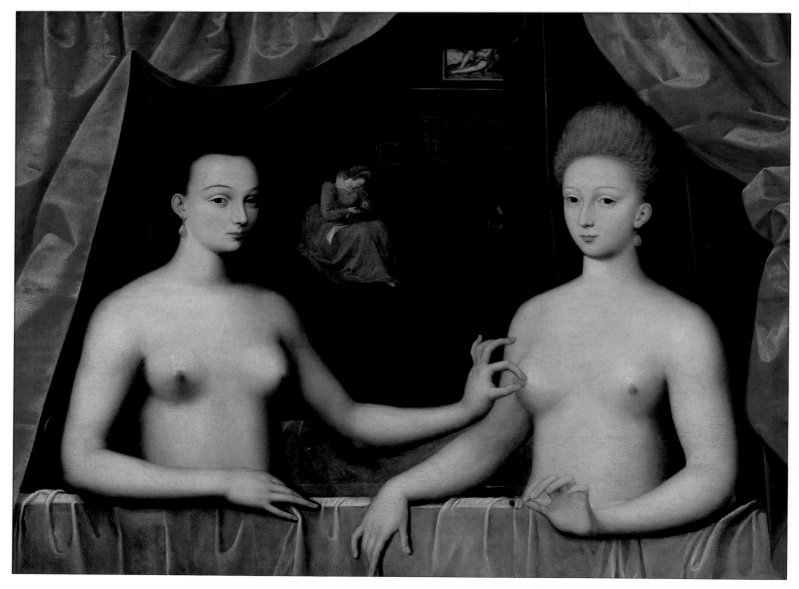

researchers think it may still be more important for humans than we think it is, but that is a difficult assertion to prove. If we listen to the anthropologists who study such things, the kiss originated as a sniff on greeting one another to detect whatever complex story body odor might have to tell about the person thus greeted.

The organization of the parts of the nervous system involved in smell is fascinating and puzzling, because they do not obey the standard rules. In the eyes and the ears special cells convert light or sound into signals that nerve cells can recognize. In the nose the first nerve fibers in the path to the brain are directly exposed to the world. The chemicals that convey the messages of odors activate them directly. More puzzling still, these primary nerve fibers (rabbits have about 100 million of them) die and regrow regularly. Then, instead of connecting to a relay station in the thalamus, the primary fibers run straight to the cortex of the brain in a region known as the olfactory bulb. Nerve fibers from this specialized part of the cortex then carry olfactory information to the primary olfactory cortex. From there, projections pass, rather puzzlingly, to the thalamus and also along other routes that connect them with the limbic system, the emotion-regulating center.

Research on the olfactory system has become quite fashionable in recent years, for the problems of figuring out how it works are exceptionally challenging. A solution seems so near at hand, but the olfactory neurons guard their ancient secrets quite craftily. It was once supposed that there were primary categories of smell, much like primary colors, with a special receptor for each. Complex odors would then be formed by combinations of the primaries. But the primary odors escaped all efforts to classify them, and today,

Smell
Perfume specialists, commonly called "noses."

Is It Real? What Our Senses Tell Us

1. Hallucinations
A young woman suffering from auditory hallucinations at the Salpêtrière hospital in Paris. From the Charcot collection.

2. Conditioned visual discrimination
Two female ducks of different species, one wild, the other domestic, select the picture of a male duck of their own species. This visual discrimination is based on conditioning techniques in which the animal is rewarded for selecting the correct picture. This work was done by Animal Behavior Enterprises.

with the increased sophistication that molecular biology has given to the study of receptors, the notion of olfactory receptors has become correspondingly complex.

Some researchers think that olfactory receptors may be as diverse as even receptors in the immune system, able to match in lock and key fashion with virtually any molecule that comes along. This view is supported by the growing realization that the olfactory system comes with little if any preprogrammed ability to identify odors and that it learns by experience. It is also clear that the olfactory system, like virtually all sensory systems, has a component that feeds information from the brain back out toward the periphery. What that information is and what role it plays in perception are far from clear. Some scientists who specialize in decoding neural signals, like Walter J. Freeman of the University of California at Berkeley, propose that the complex patterns of electrical activity generated by the olfactory bulb represent a sort of anticipation on the part of the brain about what is coming in from the nose. Others reject this interpretation of the code, but so far there is little else to go on.

One idea that motivates much research on the olfactory system today is that it might be possible to take advantage of the privileged direct access of the olfactory system to the brain to treat various mental disturbances. Or perhaps we can take advantage of the easy access of the olfactory system to the limbic system and use odors to enhance mood, improve productivity, or intervene in some other constructive way with mental performance.

Hearing is another sense that we know quite a bit about in detail but do not wholly understand. Sound is carried by vibrations of air molecules into the outer chamber of the ear. The drumhead of skin that blocks the outer ear canal vibrates in response and sets in motion a few minuscule bones inside the head. These in turn transmit the vibration to a snail-shaped apparatus appropriately named the cochlea, where the vibrations are sensed by specialized cells called hair cells. These translate the vibrations into signals recognized by the primary auditory neurons, which carry the information into the brain. The process seems straightforward, but it still conceals a number of secrets. One is the fact that we can tell the difference between two musical notes that cause movements in our cochleas so similar that we should not be able to make that discrimination.

Once the nerve signals from the cochlea have entered the ear, they travel through a series of relay stations much like all other sensory signals. We know that some of them help us determine the direction a sound comes from by detecting exceptionally small differences in the time the sound arrives at the right and left ear. Other processes seem to be involved in our ability to determine whether a sound has a low frequency, like a bass fiddle, or a high frequency, like a piccolo. What we do not understand at all, though, are how we recognize a voice as belonging to a friend or relative and how we tell the difference between pleasing and irritating sounds.

The sense of touch in many ways resembles the sense of hearing, except that the sense organs are spread out over the surface of the body instead of being packed into the inner ear. As in the case of olfaction, we have not yet solved the puzzle of how specific sensations on the skin are identified. There are a number of different types of receptors present in the skin, so early observers thought that one type identified heat or cold, another light vibration, still another heavy pressure, and so on. While this simple picture does have some truth to it, there are many features of touch that remain puzzling. The most elusive of these is the sensation of pain. What makes pain different and special?

Surprisingly, one of the basic puzzles about pain is the question of what it really is. That seems easy: pain is whatever causes you to hurt. But the pain of losing someone you love is quite different from the pain of touching a hot stove. Even if you limit yourself to considering only physical and not emotional pain, problems remain. One hundred years ago physiologists thought they had the problem solved. They said that the skin contains special receptors that respond specifically to painful stimuli, just like the eye contains receptors that respond to red or green light. These pain receptors, according to this theory, send messages to the brain over nerve fibers that constitute a pathway devoted solely to pain. There is a lot of evidence to support this theory. For instance, if a surgeon cuts the presumed pain pathway, he can often eliminate pain. But not always. And it is very difficult to identify sense organs in the skin concerned only with pain.

The opposite view holds that there are no special receptors for pain. According to this

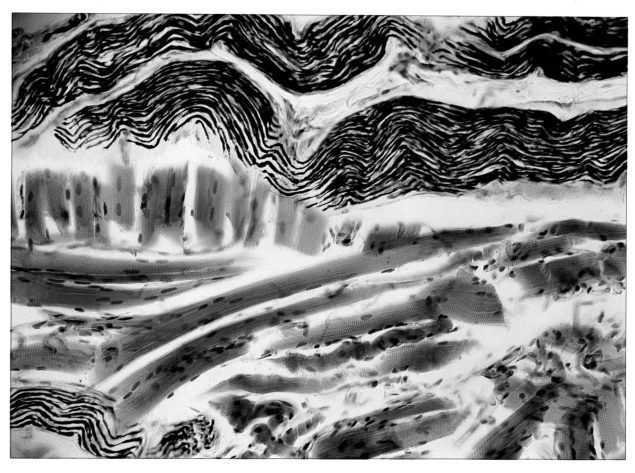

Taste
The nerve fibers that carry signals from taste-buds toward the brain are among the finest in the nervous system, with a diameter of only about a millionth of an inch. They appear as wavy black lines in this photomicrograph. The brownish structures are muscle fibers.

view, our perception of pain is simply the result of the way our brain processes a complex pattern of messages from different receptors in the skin, which, taken together, indicate that the stimulus could cause some injury. This idea that pain is all in the head has some appeal. It helps explain why people vary so widely in their response to pain, some becoming hysterical over a pin prick, while others remain oblivious even to severe gunshot wounds. It also helps to explain the removal of pain by such methods as hypnosis and acupuncture.

Modern neurophysiological research has, however, made this purely psychological explanation of pain an unsupportable one. Researchers like Edward R. Perl of the University of North Carolina at Chapel hill have clearly demonstrated the existence of pain receptors in the skin, receptors that respond unequivocally to noxious stimuli. Some respond only to mechanical stimuli that might damage the skin. Others also respond to heat and irritating chemicals. But this comfortingly simple picture no longer applies once the nerve fibers carrying pain messages enter the spinal cord and connect with neurons that course toward the brain.

Pain researchers still disagree over how to account for the many different aspects of pain and its perception. One unifying concept that many researchers find helpful is known as the gate theory of pain, proposed by Patrick D. Wall of University College, London, and Ronald Melzack of McGill University. They propose that pain messages must pass through a number of gates on their way to the brain and that these gates are controlled both by messages descending from the brain and by messages traveling toward the brain on other sensory fibers. They point out that this theory accounts for the effectiveness of transcutaneous electrical nerve stimulation and mechanical vibration in blocking pain. According to the theory, the mild electrical or mechanical stimulation activates large, fast-conducting nerve fibers with signals that travel into the spinal cord, where they block the pain messages traveling along smaller, slow-conducting fibers. Researchers also point out that the many peptides in the spinal cord, like pain-killing enkephalin, could play an important role in the gate control system.

Patrick Wall sums up the impact of the gate theory of pain on the issue of mind and brain as follows:

A study of the mechanisms of the sensory transmission pathways shows that they are dominated by fast gate control mechanisms and slow connectivity controls. These positively select those events which are positively allowed to influence the course of the behaviour. We as conscious persons are not therefore trapped inside our heads relying on the rigid mechanism of a steady flow of all available information. On the contrary we sense events in a series of sensory worlds created by our exploration and selected from all possible inputs. The exploration of the outer world generates a sensory input, and that input is selected from the entry point by laws related to the needs of behaviour in progress. This view of sensation unites mind and body into a single subtle entity rather than the two-stage view of dualists, where a mechanism without quality is followed by a mind which is gloriously clever but dependent on a stupid mechanism.[1]

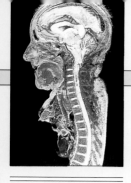

The Neural Basis of Olfaction

by Eric Barrington Keverne, Department of Anatomy, University of Cambridge

The sense of smell (olfaction) has never received the kind of investigative attention that scientists have committed to the other senses, especially vision. Yet the use of chemical communication in many aspects of mammalian behavior is undoubtedly important for survival. Behaviors such as feeding, courtship, maternal behavior, and defensive behavior are all heavily dependent on olfactory signals for their initiation and functioning.

Although it is generally accepted that most mammals use all of their senses to assess their environment, some clearly rely on olfactory information more than others, especially in reproduction. Thus, not only do mice identify the sex of an individual by its odor, but their physiological reproductive state is determined largely by chemical cues (pheromones). Mice clearly fit into the category of macrosmatic mammals (possessing a highly developed sense of smell). On the other hand, primates, including man, have all of their senses well developed and with the evolutionary enlargement of the neocortex have the capacity to assimilate and integrate information rapidly from a number of sensory channels simultaneously. Behavior does not come under the obligatory domination of any one sense; all senses are used to derive maximal information from the environment. Neuroanatomists have traditionally classified primates as microsmatic. What they mean by this term is that a relatively small area of the brain is given over to the olfactory sense, and the peripheral receptors are fewer in number than in a macrosmatic species, such as the mouse or rabbit. Indeed, a cursory look at the olfactory centers of the brains of a mouse and a monkey provides convincing anatomical contrasts. In the mouse the olfactory bulbs occupy a large part of the cranial cavity, whereas the olfactory bulbs of a monkey are barely visible and are totally subordinated by the neocortex. Similarly, macrosmatic species have a larger number of receptor cells in the olfactory mucosa than are found in a monkey (contrast the rabbit's 100 million receptors with the monkey's 10 million).

Now, if receipt and interpretation of the olfactory message relied solely on these anatomical features, primates could certainly be considered microsmatic. However, two important findings have invalidated this method of classification. The first came when recordings from single olfactory receptors became possible and it was discovered that such receptors are not narrowly tuned to particular odors but can respond to a number of different odors. This favored the hypothesis that a pattern of receptor firings was the means of coding odor cues, rather than a labeled line of a single receptor type for each distinct odor. The second finding was that the neural projections of the olfactory system have access to the neocortex via the thalamus. The fact that olfactory information is coded as a pattern requiring a higher order of recognition means that animals with a greater neural backup have the potential to recognize more patterns and hence more odors. Whether or not this potential is fulfilled is open to question, but simply counting neurons fails to account for how the neurons function and is no indicator of perceptual capabilities.

Most mammals, except higher primates (including man), possess two types of olfactory receptors. These are the chemosensory cells in the olfactory mucosa of the nasal cavity, which project to the olfactory bulb, and the receptors of the vomeronasal organ, which project to the accessory olfactory bulb. The central projections of the main and accessory olfactory bulbs also remain distinct, the latter projecting directly to the amygdala in the limbic brain, and the former to the pyriform cortex. The pyriform cortex is in turn linked to the neocortex via the thalamus and also has multiple projections into the limbic brain (septum, amygdala, hypothalamus), while the accessory olfactory system projects only to the amygdala and from here principally to those parts of limbic brain concerned with motivated behaviors and neuroendocrine function. One can best ascribe a meaningful distinction between these dual olfactory projections in the context of reproduction. While the main olfactory system provokes sexual arousal, the accessory system primes the female for ovulation and implantation and increases the secretion of gonadal testosterone in the male. The diffuse cortical projections allow for a degree of plasticity in the behavioral response to odor cues, but the direct limbic projection induces an invariant neuroendocrine response.

Primates' loss of this priming accessory olfactory system is part of the evolutionary process by which man has become emancipated from his physical environment. For example, man is not a seasonal, photoperiodic breeder, nor is ovulation reflexly induced by tactile stimulation in man, nor is there evidence for pheromonal priming in man. While this is one obvious level of hierarchical processing that distinguishes man from other mammals, even in man the main olfactory system has pronounced access to the emotional brain. It might even be argued that this strong association with the limbic brain and the paucity of connections with neocortex, especially association cortex, provides the olfactory sense with a means of influencing our better judgment. Witness the power of odors added to detergents or soaps, the fresh-linen smell of new clothes, to persuade one to buy. Even fruit and vegetables may have their odors enhanced to promote their appeal. Moreover, the success of the perfume industry is based not on the rational judgment of buyers, but on rational assessment of the hedonistic, emotional behavior of buyers.

Olfactory neurons undergo a process of continuous turnover that takes about 30 days. Since each receptor is not specifically coded for a given odor but contributes to a pattern of firing across many receptors, this pattern must also undergo continuous change. There is, however, considerable convergence of receptors in the synaptic glomerulus of the olfactory bulb (1,000:1), and hence considerable redundancy in the system. The extent to which this convergence overcomes the problems created by turnover is not known, but any change in the distribution of receptor types could clearly have consequences for smell memory. We have no ability to call up smells and rehearse them in our mind's eye, as we have for visual imagery, and while smells can recall visual images, the reverse is not possible. Perhaps this is why we have very poor smell memory: complex smells may only endure the active life of a given population of receptors, and without repeated exposures over time, their memory may fade as the receptor population changes.

Smell is a very personal sense in that perceptions may vary widely in the population. Even pure chemicals, as opposed to the more complex molecular mixtures, can be perceived differently. The steroid androstenedione may smell sweet and musky to some and urinary and unpleasant to others. It has no smell at all to approximately 20 percent of the population. When more than one odor is presented simultaneously, we are often unable to identify the components of the mixture, just as we are unable to distinguish the component wavelengths of white light. This is because the pattern of receptor firing generated by each individual odor is lost in the total pattern produced by the odor complex. It is true that skilled perfumers can analyze odor complexes, but it is likely that they have learned to recognize the patterns that given odors produce when mixed, rather than that they have any memory capacity to separate them analytically.

In man the appreciation of such olfactory codes appears to be mainly hedonistic, an important adjunct to the appreciation of wines and perfumes, though learned chemosensory discriminations do allow us to identify many of the artificial compounds (some toxic or explosive) that have been introduced into the environment. Thus, the development of the neocortex may have freed man from dependence on the chemical senses for survival, but this same neocortical development has permitted the hedonistic and cognitive appreciations of the trained nose.

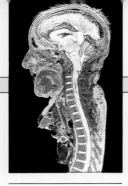

There Is More to the Smell of Pigs Than Meets the Nose

by W. D. Booth, Agricultural and Food Research Council, Institute of Animal Physiology, Cambridge, England

Few people need reminding that the pig is a smelly animal. When one looks more closely into the reasons behind this centuries-old association of pigs with unpleasant odors, some interesting aspects about the biology of this animal begin to emerge. We share as much responsibility as the pig in the creation of the more obvious malodors coming from the pig pen, since we have compelled this intelligent beast to exist in the relative confinement of our intensive farming systems. The malodors arising from the pig's excrement are concentrated not only in the corner of a pen but also on the animal as it lies in proximity to its waste. Yet amid this plethora of smells are some odors produced by special glands in the pig that are important in olfactory communication indicating individual or sexual status.

The pig, like most foraging and hunting mammals, has a well developed sense of smell, in contrast to the relatively diminutive olfactory system found in man. The pig's acute sense of smell is exploited by man when he trains pigs to search for truffles. It is perhaps no coincidence, as we shall see later, that there is apparently an odorous compound in truffles that is also present in the pig and man.

A useful experimental approach for determining the importance of the sense of smell is to remove it (anosmia) by surgically removing the olfactory bulbs from the brain. Pigs are normally very sensitive to unfamiliar pigs, but after olfactory bulbectomy they more readily accept strangers during feeding and, in the case of the anosmic sow, will foster another sow's piglets. In female pigs, olfactory bulbectomy delays puberty, and in mature animals it disrupts the regular three-week estrous cycle for long periods during the summer.

There seems to be some justification for considering the male pig, or boar, to be a bit of a chauvinist. His sexual maturation is not unduly affected by olfactory bulbectomy. Furthermore, his gonads produce not only sperm and male hormones but also one of nature's aphrodisiacs, the musk-smelling steroidal pheromones 3-alpha-androstenol and 5-alpha-androstenone. These substances are found only in small amounts in sows. The pheromones are concentrated in the boar's fat and skin glands and particularly in the sexually dimorphic submaxillary salivary glands and saliva by a specific binding protein. When a boar is aroused in the presence of estrous pigs or unfamiliar boars, he champs large amounts of foamy saliva. This amplifies the release of the pheromones, which remind the estrous pig that she should be in a receptive state for the male or signal another male that someone is challenging his authority. Exposing immature female pigs to boar pheromones and other boar stimuli for at least 20 minutes a day will promote puberty in these pigs.

Little is known about the mechanisms in the pig's brain that act after an odor impinges on the olfactory epithelium. One can speculate from studies in rodents and sheep that neural connections between the olfactory bulbs and a primary integrative center in the brain, the hypothalamus, are also important for normal hormone secretions and behavior in the pig.

What is the link between truffles, pigs, and men other than the thought of food? By digging below the surface we find food for thought. Apparently, the boar pheromone 3-alpha-androstenol is present in the truffle *Tuber melanosporum,* which possibly provides a reason why pigs seek truffles. But perhaps this food-chain link goes further, since 3-alpha-androstenol also occurs in men (and a little in women). Is this where our liking for both truffles and pig meat comes from? Yet too much boar pheromone in the fat of boars, particularly 5-alpha-androstenone, taints the meat. The most challenging food for thought is the significance of the boar pheromones in the human. Although most women dislike the smell of boar pheromones, they have been added to cosmetic preparations in the wake of the use of 5-alpha-androstenone in commercial aerosols for detecting estrus in pigs. But no effects in humans comparable to those found in pigs have yet been demonstrated in well-controlled experiments. Our relatively insensitive sense of smell will safeguard us, one hopes, against attempts to manipulate our behavior through our olfactory systems.

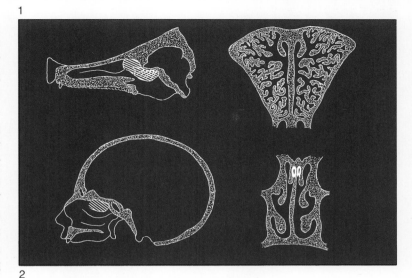

1. A comparison of olfactory epithelium in the pig and the human
Left upper and lower drawings are sections through the skull of a pig and a human respectively. Right upper and lower drawings are enlarged sections at right angles through the hatched areas in the corresponding skulls. The olfactory epithelium lines the whole labyrinth in the pig, while in the human it is confined to the small U-shaped areas shown.

2. Pheromones
In the animal world, pheromones, chemical messages that disperse through the air and are smelled by other animals, play an important role in behavior. Pheromones play a particularly important role in attracting members of the opposite sex. Here a mature boar secretes a foamy saliva laden with pheromones.

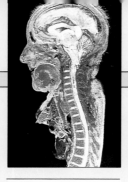

Hearing: A Collaboration of Brain and Ears

by Nelson Yuan-shen Kiang, director, Eaton-Peabody Laboratory, Massachusetts Eye and Ear Infirmary

The human auditory system is the product of a long evolutionary line of structures that can detect mechanical disturbances in the environment. Even unicellular organisms respond to mechanical stimuli. Early in the evolution of multicellular organisms, distinct cells appeared that were especially sensitive. These cells were connected with muscle or other effector cells either directly or through intermediate cells (neurons) specialized for transmitting information sometimes over long distances by means of propagated activity in extended processes. The basic elements of an auditory system can already be identified in primitive invertebrates. The essential elements are: (1) specialized sensory cells that release chemical transmitters when they are mechanically stimulated; (2) primary neurons that discharge when stimulated by the transmitter released by the sensory cells; (3) chains of neurons whose activities are influenced by the primary neurons; and (4) motor neurons or other effector neurons.

The foregoing description holds for any of the so-called hair-cell systems. Hair-cell systems include lateral-line organs in fish and amphibians for sensing water movement, vestibular organs in vertebrates for detecting head motion and gravity orientation, and sound detectors in vertebrates and also invertebrates like many insects, which exhibit a wide variety of adaptations for detecting air-borne sounds. Thus, the human auditory system is but a special case of a mechanoreceptor system, one that has assumed special importance through our dependence on air-borne sounds as carriers of information between individuals. It is hardly an exaggeration to say that speech communication has been vital to the development of group relations, language, and culture in human societies as we know them.

The key to understanding the mammalian auditory system is to trace the relevant signals through successive stages of the system. As indicated above, a critical stage is the sensory cell, which performs the important function of transduction. Consider a hair cell with stiff stereocilia (hairs) extending from the top (see figure). These stereocilia are bent relative to their resting upright stance by mechanical stimuli. In some as yet unknown way, this bending opens ion channels somewhere near the top of the cell. The electrical changes in these cells follow the mechanical or sound stimulus so faithfully that they are called "microphonic potentials," because they resemble the electric waveforms produced by a microphone in response to sound. The cellular changes in electrical potentials act as signals to the chemical-release mechanisms located at the opposite end of the cell. The as yet unidentified chemical transmitters are released to act upon the membranes of the nerve endings that lie under the base of the hair cell so that when the stereocilia are bent in one direction, the fibers innervating that hair cell have an increased probability of firing and when the stereocilia are bent in the opposite direction, the probability of firing decreases.

Hair cells are an essential component of auditory systems, but great modifications have been made in the settings within which they operate. At the front end a special sense organ, the ear, evolved to deliver sounds to hair cells more effectively. The most visible part of the ears, the external ears, are laterally placed, paired structures shaped to funnel sound into the

The mammalian auditory system

130

middle ear. These external ears often have sculptured contours that have the effect of absorbing certain frequencies. Because our external ears are asymmetrical in a front-back direction, we can distinguish complex sound sources in front of us from those in back. This is possible even when the sources are in the midline, which makes relative intensities and times of arrival at the two ears the same. Many mammals can turn their ears to scan their environment or track a moving sound source.

At the level of the middle ear, the definitive feature for mammals is a chain of three ossicles: the malleus (hammer), incus (anvil), and stapes (stirrup), connected in series. This chain serves to transmit sound to the inner ear. The mammalian middle ear also has two middle ear muscles, the tensor tympani, attached to the malleus, and the stapedius, attached to the stapes. These muscles not only contract reflexly in response to loud sounds but can also contract before or during vocalization, chewing, or generalized startle responses. In general, mammals can detect high-frequency sounds better than other vertebrates. In keeping with this capability, the overall effect of contractions of middle ear muscles is to reduce transmission of low-frequency components into the inner ear without a comparable reduction of high-frequency components. Most background environmental noise is low-frequency, so the middle ear muscles may be effective in keeping the ear sensitive to high-frequency sounds by reducing the masking effects of background noise. There may also be a protective effect: the transmission of high-level sounds, which might otherwise damage the ear, is reduced.

In the inner ear the cochlea, an intricate structure, has evolved to maximize the sensitivity of hair cells by making them more sharply tuned to specific frequency ranges. The hair cells are held within a complex of supporting cells sitting atop the basilar membrane, which is surrounded on both sides by fluid-filled channels. This membrane, running the length of the cochlea, is taut and narrow at one end (the base) and flaccid and wide at the other end (the apex). The fluids in the cochlear spaces differ in content, and there are large differences in electrical potentials between some of the cochlear fluid compartments. These specializations are thought to increase sensitivity to low-level sounds. The dimensions of individual hair cells vary systematically from base to apex. All of these morphological features combine to increase sensitivity and sharpness of tuning in individual hair cells and their attached neurons. The more apical cells are more sensitive to low frequencies and the more basal cells are more sensitive to high frequencies. Complex sounds with a mixture of frequency components can be sorted out by the cochlea, since individual hair cells are tuned to specific ranges of frequencies. This mechanical analysis of sound not only provides exquisite sensitivity; it also provides a basis for generating distinctive space-time patterns of hair-cell activity determined by the frequency content of a sound stimuli. Because the innervation of inner hair cells is so punctuate, sound creates distinctive patterns of activity in the array of neurons that make up the auditory nerve, providing the central nervous system with information about the sound stimulus.

The auditory nerve carries signals that are at once the output of the cochlea and the input to the brain. The signals are in the form of all-or-none electric pulses (action potentials) that are propagated without loss of amplitude over long distances. Thus, neural signals can be transmitted over long distances, combined, compared, and manipulated to achieve what is generally described as functional integrations. The selection of appropriate motor responses is based on the meaningfulness of the acoustic cues as constructed from the discharge patterns carried in the auditory nerve and interpreted by the brain.

The mechanisms by which neural signals are centrally analyzed depend in part on an orderly spatial representation of signals. At each neural stage there are arrays of neurons arranged in some topographic order. Neurons are arranged according to input projections ultimately traceable to the longitudinal location of the hair cells innervated. More specifically, auditory nerve fibers enter the cochlear nucleus, the first place where activity patterns can be modified, transformed, and redistributed to more central locations in the brain. Each auditory nerve fiber distributes its endings in an orderly manner. The cochlear nucleus is clearly cochleotopically organized, but cells maximally sensitive to the same frequencies can differ markedly in other response characteristics. In particular, the time patterns of discharges may be drastically changed from those of auditory-nerve fibers. For example, some cells will only fire to the onset of sounds, whereas all auditory-nerve fibers will respond as long as the stimulus is present, provided the appropriate frequency components are present. The activity of some cells of the cochlear nucleus will accurately reflect the cycle-by-cycle timing of the stimulus, whereas others will not be so temporally coherent. Some cells will have much larger dynamic ranges for a rate of discharge than those of auditory nerve fibers. Cells of like kinds tend to have their output projections grouped together to form fiber bundles that travel to specific nuclei at various places in the brain stem. One sees, then, that the operating characteristics at virtually every part of the system can be modified by activity of the central nervous system. The roles of such efferent control might be selective attention or protecting the ear from overstimulation.

Disorders of Hearing

Having described the general nature of mammalian auditory systems, we may now discuss how such knowledge helps us to understand faulty hearing. The idea that the hair cell makes up a key stage in the system has immediate import. It is difficult to produce total deafness without involving the cochlea or the cochlear nucleus. The reason is that defects at more peripheral stages influence only sound transmission, which can rarely be entirely abolished. Amplification of sound or surgical alteration of sound-transmission paths can usually improve hearing defects due to blockage of sound. Similarly, defects of a neurological nature rarely affect all the auditory pathways once the signal passes beyond the cochlear nucleus. Central tumors, vascular defects, or degenerative conditions that are so widespread as to affect *all* aspects of hearing would most likely be fatal.

Defects at the level of the hair cell or at more central levels produce sensorineural hearing losses and are more difficult to treat than conductive losses. Sounds of very high intensity, ototoxic drugs, vascular insults, infections, and genetic defects are a few of the conditions that can damage hair cells. Under some of these conditions, viable spiral-ganglion neurons remain even after elimination of hair cells. Recent attempts to produce useful hearing that use electrical stimulation to bypass nonfunctioning cochleas have been encouraging. This type of prosthesis (a cochlear implant) serves at the very least to provide cues that supplement those available from lipreading. Voicing cues, for example, depend upon factors that are not visually detectable but are easily delivered by electric stimulation. With available devices some patients are able to discriminate words in open-set word lists and conduct free-running conversations reasonably well, particularly if lipreading cues are also available. At present, those of the profoundly deaf community who have learned to sign are generally reluctant to accept such a device-oriented solution, since they have already made adjustments in their life styles. However, for congenitally deaf children who have difficulty developing language skills, the possible advantages of cochlear implants done in the early years should be seriously examined. As implant devices improve, the balance seems certain to swing toward more general use of such devices. Being able to monitor the acoustic environment, to join the hearing community, and to produce more adequate speech are major possible benefits to be gained through successful implants. Knowledge of how specific acoustic cues are coded in normal hearing forms a basis for improving future implants.

Modern methods of recording electrical activity from the ear and brain with surface electrodes are enabling us to diagnose hearing problems in infants as well as adults and to determine whether the defects are of ear or brain origin. There is hope that, as these tests improve and are supplemented by modern imaging methods, many of the as yet unexplored individual variations in auditory-processing capabilities will be attributable to specific problems of organic origins. Ultimately, such studies will add impetus to studies of the interactive roles played by the brain and the ear in processing auditory information.

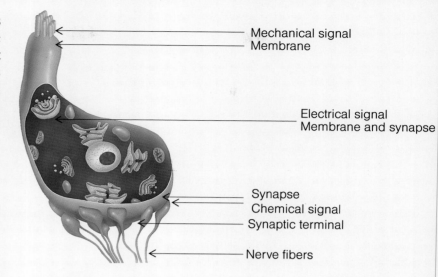

Hair cell
A longitudinal cross section through a representative sensory cell (based on mammalian inner hair cells) showing stereocilia at the top and chemical-release machinery at the bottom. Also shown are afferent nerve fibers that end on the base of the hair cell. Motion of the stereocilia depolarizes the cell and leads to the release of transmitter substances that depolarize the auditory-nerve endings and thereby cause action potentials.

10
Who's in Charge?
The Control of Movement

How does the brain initiate and coordinate the action of muscles? Research on the nervous system early in the twentieth century was surprisingly successful in revealing how reflexes work and raised hopes that this question soon would be answered. Tap the knee and a barrage of nerve impulses travels into the spinal cord, jumps the synapse to a motor neuron (one that activates a muscle), and jerks the leg forward. With the mechanism of reflexes firmly in hand, scientists proposed that all movements, even such complex movements as playing the piano, are but a cascade of interlocked reflexes. Today we know that this idea is simply not correct, but only slowly as researchers learn to ask the right questions is the true nature of motor control coming to light.

Unlike sensory systems, which can be fairly neatly compartmentalized into vision, hearing, and so on, motor systems are not easily broken down into component parts. It is also difficult to sort out what the brain is controlling when it directs movements like picking up a pencil or throwing a ball. Is the goal of the brain the final position of the hand at the pencil, and to achieve that goal does the brain constantly update the path of the hand as it travels toward the pencil as a computer guides a guided missile? Or is there a single command that instructs the arm to swing in an arc and release the ball at a certain position? Both of these pictures seem to be true to some degree, and new kinds of experiments are helping us to discover how the brain makes our muscles work.

Early in the second half of the nineteenth century the prevailing view of neurologists was that the highly evolved human cortex was too lofty to deal with such mundane matters as controlling the activity of muscles. It supposedly attended only to thinking and other high-level functions. Control of the muscles was relegated to the brain stem and spinal cord. British neurologist Hughlings Jackson became quite unpopular with his colleagues for his radical proposal that the cerebral cortex contains regions that have direct control over the contraction of muscles. He further postulated that these cortical regions are organized to activate not just single muscles but coordinated groups of muscles required to produce basic movements. For example, a cortical region that produces hand movements would presumably have outputs that activate the arm, shoulder, and trunk movements that are normally involved in hand movement. By implication this second hypothesis says that because a single muscle may be involved in a wide variety of movements, it should be represented by widespread or possibly multiple regions of the motor cortex.

It was not long until the German physicians Gustav Fritsch and Julius Hitzig vindicated the first of Jackson's proposals by demonstrating that electrical stimulation of the exposed cortex of a dog evokes movements of the body. The second of Jackson's proposals, the organization of the cortex according to movements rather than muscles, was to remain under dispute for some years. Extensive efforts ensued to map the motor cortex of anesthetized animals by applying electrical stimulation point by point and observing the resulting movement. This work, which culminated with the unsurpassed efforts of Clinton Woolsey at the University of Wisconsin in the 1950s, all seemed to suggest that each cortical location activated a particular muscle. Woolsey's experiments with monkeys made it possible to draw a distorted picture of the animal on the motor cortex showing which brain region activates which muscles. Such a map suggests that the body is represented within the motor cortex in a continuous fashion, with only one area for each body part, an arrangement quite contrary to Jackson's second proposal.

The development of increasingly sophisticated technologies for electrical stimulation of the brain in the decades following Wool-

Judo
Judo was created as a sport late in the nineteenth century by drawing upon several systems of bare-handed fighting, such as jujitsu, which in turn were inspired by ancient Japanese mythology. The sport has many complicated rules, but a principal objective is to throw the opponent with controlled force so that both his feet leave the mat or to hold him under control on his back for 30 seconds. A judo match is a test of skill and control, not of endurance, and is supposed to exercise both body and mind to the fullest extent.

Who's in Charge? The Control of Movement

1-3. Karate and aikido
Unlike judo, karate, meaning "empty hand," originated as an art of self-defense, not as a sport. Its origins date back to the sixth century, but it was perfected on the island of Okinawa during the seventeenth century, when the occupying Japanese forbade the bearing of arms. Karate involves hardening the extremities by repeatedly striking walls and other hard surfaces until the hands and feet themselves become capable of maiming an opponent (1 and 2). The martial art of aikido (3) originated during the time of the samurai, the warrior aristocracy, in fourteenth-century Japan. The basic principle of this method of self-defense is to not resist and to use the opponent's own momentum as a weapon. The speed of movement can be dazzling.

sey's work enabled many investigators to pursue the question of muscle versus movement and produced conflicting and confusing results. In 1978 H. C. Kwan and his colleagues at the University of Toronto broke new ground by inserting tiny wires into the arm area of the cortex of unanesthetized monkeys, using minute currents to provide highly localized stimulation. Instead of observing muscle twitches, they observed the relationship of the evoked movements to the major joints of the hand and arm. Their unexpected results suggested an organization altogether different from Woolsey's map. What appeared was a pattern of concentric rings resembling a target. Stimulation of the bull's eye evoked movement of the fingers. Stimulating the wire in the surrounding ring led to movements of the wrist; stimulating movement up the next ring, the elbow; and stimulating movements of the outer ring, the shoulder. Here was evidence that while Woolsey's map may be anatomically correct, its functional implications may have been misread.

The current view emerging from the work of Donald Humphrey at Emory University is that individual muscles are widely represented over the motor cortex and that several muscles are represented at any one location. These muscles are the ones required to execute a simple movement about a joint. When you flex your fingers, for example, you must contract your wrist extensor muscle to prevent your wrist from flexing, since the finger flexor muscles also flex the wrist. This is quite in keeping with Jackson's second proposal and suggests that relatively complex movements can be brought about by a simple activation of a limited cortical region. These results, however, were obtained by unnatural electrical stimulations of the cortex. Much research remains to be done to determine whether naturally occurring coordinated movements are brought about by activation of limited areas of the motor cortex by nerve signals from other brain regions.

The discussion of consciousness in chapter 1 presents evidence that there is electrical activity in the brain that preceeds a conscious decision to move. This raises the possibility that the brain does something to set the stage for movement, perhaps formulating some sort of plan to carry out a movement that involves many different muscles acting in just the right relationship with one another to

Who's in Charge? The Control of Movement

bring about a desired objective, like picking up a pencil or swerving to avoid running into another car. How can we pry into this process to see what is going on?

One approach used by Apostolos P. Georgopoulos at the Johns Hopkins University involves teaching monkeys to reach for a light that goes on somewhere on the surface of a sphere that surrounds them. He recorded electrical signals from nerve cells in the regions of their cerebral cortexes directly responsible for initiating muscle contractions. Some of these cells do indeed become active before the monkey reaches toward a light. As it turns out, different cells become very active when the monkey reaches in a specific direction and are less active as the angle of reach extends away

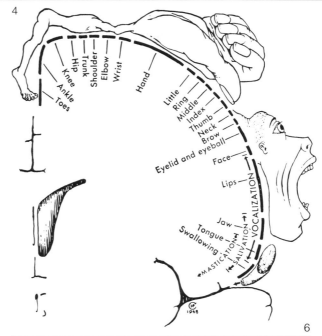

4. The movement homunculus
The cartoon man (homunculus) sketched against a section through the human motor cortex illustrates the relative dexterity of different body parts. The large size of the hands and the area of the mouth reflect their great dexterity in comparison to that of the toes and ears.

5. Restoring lost movement
In this experiment a paraplegic patient takes a few steps with the aid of electrical impulses delivered to his leg muscles. The pack on his belt generates the impulses under command from a computer.

6. Dance, or the harmony of movement
Master of the control of movement, the legendary Russian ballet star Vaslav Nijinski (1890-1950) was regarded as the supreme male dancer of modern times. His phenomenal technique enabled him to leap to seemingly impossible heights and to appear to remain suspended in air. In 1917 his brilliant career was cut short by mental illness of an unknown nature.

from the preferred direction. Georgopoulos drew a three-dimensional graph and placed an arrow on it that points in the direction that gives the maximum response of each nerve cell and has a length that represents the strength of the activity of the cell. He found that averaging all the arrows together gives one resulting arrow that points in just the direction the monkey is going to move its arm. So the brain seems to compute the average activity of a set of cells to determine the direction of a forthcoming movement.

Some related experiments, again with monkeys trained in a movement task, were conducted at the Rockefeller University by Hiroshi Asanuma. His monkeys had to reach for a food pellet brought toward them in a hole in a rotating wheel. Like Georgopoulos, Asanuma also recorded electrical activity from cells in the primary motor cortex. He noted that columns of cells all send their signals to one local muscle region and also receive information back from that same region in a sort of closed loop. He too found that about 10 percent of the cells start delivering signals a fraction of a second before movement starts. At the same time, very low levels of electrical activity were recorded from the muscle region, which suggests that the muscle was being prepared for action. But when Asanuma blocked the pathway carrying information from muscle to brain, the prior electrical activity disappears and the monkey can no longer perform the task.

As in most areas of brain research, many investigations of how the nervous system brings about complex patterns of movement have looked to the movements of simpler creatures with the hope of finding simpler answers. The laboratory study of how insects fly and how newts walk has indeed told us something about movement control, but, as seems to happen all too often in brain research, the answer posed still more questions. In the case of rather complex movements made by simple organisms, the not-so-surprising solution to the problem is that the animal's nervous system provides some sort of program, something like a computer program. In this case the program directs the contraction and relaxation of the creature's muscles. But what is this program, and how was it written?

A motor program, a program for movement, must specify which muscle is to contract or relax, when it is to perform such a function, and the proper degree of contraction or relaxation. It is, in a way, similar to a musical score that calls for each instrument to begin and cease playing at a precise time and at a precise level to perform the musical piece. Such programs first came to light in the 1960s in research on the movements of locusts and crayfish. The movements of these creatures surprised researchers at the time by appearing to be recorded in the nervous system as if they were on a cassette tape.

Some of the most elegant experiments were done by Donald M. Wilson (now deceased) at Stanford University. He demonstrated that the locust's two pairs of wings move in response to bursts of electrical impulses produced by a group of nerve cells in a structure called the thoracic ganglion. About 20 nerve fibers run from this ganglion to each wing, and the bursts of impulses are timed so that muscles pulling the wings up and down are alternately activated. The pattern of signals to the wings persists even if the thoracic ganglion is isolated from signals from all other parts of the insect's nervous system; the cassette continues to play. Precisely how even this simple motor program is generated continues to elude researchers to this day.

It seems quite likely that humans too have motor programs built into their nervous systems. One practical application might be the ability somehow to tap into such a motor program for walking in the spinal cord and activate it electrically to help patients with damaged spinal cords to walk. That problem is not easily solved, of course. Even the locust needs information about wind speed, lift forces, and the location of the horizon to govern precisely the flight pattern it follows. The same would presumably be true for a program for walking in the human nervous system. And since information from the eyes and the organs of balance in the head is cut off from the hypothetical motor programs in the spinal cords of such patients, some arrangement would have to be made to provide appropriate stimulation to postural muscles as well.

Much of what we actually know about the control of movement in humans comes from clinical observations of individuals who have suffered damage to some part of the nervous system. Regrettably, most of what they tell us confirms the opinion that the control of movement is a terribly complex process. We

Who's in Charge? The Control of Movement

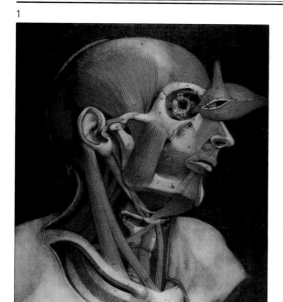

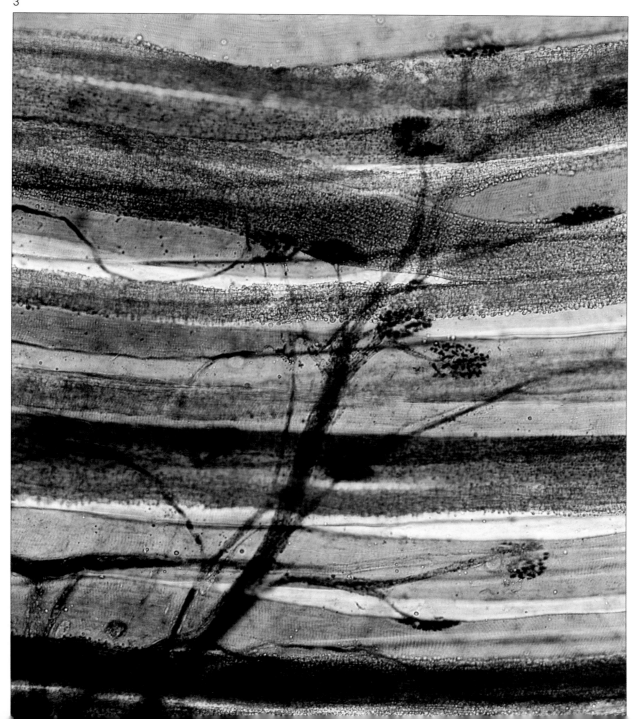

1. Écorché
An écorché, a French word meaning flayed, is an anatomical drawing in which the skin is stripped away for the study of muscles. This engraving, showing the muscles of the head, is by J. F. Gautier d'Agoty, 1748.

2. A monkey affected by the drug MPTP
In July 1982 a 42-year-old drug addict was hospitalized in San Jose, California. He showed the symptoms of Parkinson's disease, which puzzled the doctors, as he was relatively young. It did not take long to establish that the cause was an impurity in the street drug he had been taking. The impurity, MPTP (N-methylphenyltetrahydropyridine), leads to the destruction of the same brain cells that degenerate in Parkinson's disease. The accidental discovery of MPTP now makes it possible to study Parkinson's disease in animals and to develop drugs for its treatment.

3. Nerve-muscle connections
Nerve fibers, the black lines sweeping up from the lower left, make contact with muscle cells, the horizontal brown bundles, at specialized junctions that appear as small black dots. An electrical signal arriving on a nerve fiber causes the release of the neurotransmitter acetylcholine into the microscopic synaptic gap between the nerve fiber and the muscle. Arrival of the neurotransmitter causes a change in the electrical properties of the muscle cell, which in turn causes its contraction.

need all the research tools at our disposal to unravel it. And we need to learn how to ask crucial questions that can be answered with the tools available to us. This is clearly one of the darkest corners of brain research, but a glimpse into the shadows may reveal a few general principles (which may yet prove incorrect) about the brain's control of our muscles.

Direct clinical investigation of what is apparently the highest control center for movement, the motor cortex, really began with the work of pioneering Canadian neurosurgeon Wilder Penfield. He used mild electric currents to stimulate different regions of the brains of patients about to undergo brain surgery. The patients had only local anesthesia and were wide awake at the time. The electrical activation of regions of the motor cortex caused the patients to move some part of the body on the side opposite the stimulated cortex. (The outgoing nerve fibers cross to the opposite side). But surprisingly, the movement occurred without the patient having any sense of having played any role in initiating it. The question of whether voluntary movements actually originate from activity in the motor cortex has not yet been satisfactorily answered. Cats that have been experimentally deprived of the motor cortex can go about their daily business in a seemingly perfectly satisfactory manner. They can even give birth to and care for kittens. The motor cortex must play an important role in movement, but lower nerve centers are clearly capable of doing quite a few things on their own.

A real problem in the study of the control of movement is the fact that the control systems in the nervous system seem to be made up of layer upon layer of feedback loops. This means that movement of a muscle sends signals back to the nervous system to tell it what is happening. These messages then cause the nervous system to change the signal it sends to the muscle. It is, in a way, something like the loop between a furnace and a thermostat. When the temperature drops below the setting of the thermostat, the thermostat turns on the furnace. The furnace heats the room until it rises above the setting, and the thermostat turns the furnace off until the temperature again drops too low.

A loop like this exists if you try to place your finger on a moving spot on a television screen. You move your hand rapidly toward the screen, and as your eyes tell your brain that your finger is approaching the dot, your brain slows the muscle movements, adjusting them until the distance between the dot and your finger has vanished. Then your eyes provide information needed to move the finger to keep it on the dot. But within this large loop between the eyes and the finger-moving muscles there are loops operating within loops within loops of control systems. It becomes so complex that the only hope to understand what is happening in the nervous system to control movement seems to come from the tools of control-system engineering. Such tools are large, powerful, and complex computer models capable of analyzing systems with large numbers of interacting components, systems like the economy of a country.

The major components of the human motor system are rather easy to identify and have been studied extensively. Yet the details of how the individual components function and how they interact to provide us with the capability of moving purposefully remain frustratingly elusive. The motor cortex seems the obvious source for the origins of movement, but research shows that this idea is too simplistic to explain what happens. Indeed, the initiation of movement depends on a motivation to move, and it is the limbic system of the brain that appears to serve this function. Measurements of cerebral blood flow in humans show that cortical regions in front of the motor cortex become active before a voluntary movement is made. The motor cortex becomes active only when the movement is actually performed.

In addition to motivation, movement requires an internal program for its generation, a variety of sensory triggers for its initiation, and guidance provided by information from the sense organs. It is the cerebellum that plays a major role of providing sensory guidance for movement. Other major parts of the brain's motor control system known as the basal ganglia seem to be involved in the selection and production of primitive subcomponents that provide the basic structure of movements. And all these parts interact with one another during movement to produce the end result, the smooth purposeful movement that appears so simple.

Years of painstaking work have revealed that there is a kind of functional subdivision of

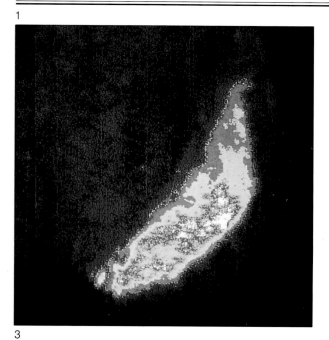
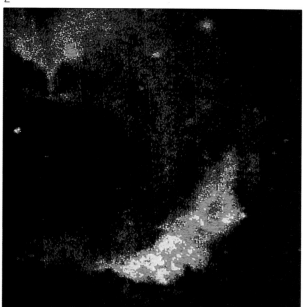
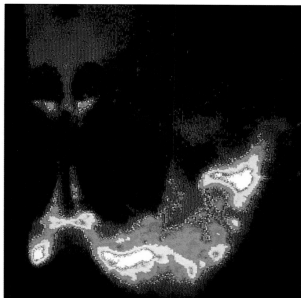
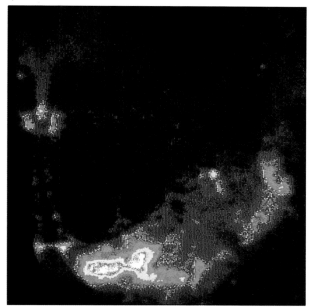

1-2. Age related loss of neurotransmitter receptors
Aging is accompanied by a decline in a number of physiological and biochemical capabilities. For example, the number of neurons and the concentrations of some neurotransmitters and their receptors decline with age. This is illustrated here by a marked decrease in the number of dopamine receptors in the human substantia nigra, the area of the midbrain occupied by neurons that use the neurotransmitter dopamine. A radioactive substance that binds to the receptors was applied to thin sections of the brain, which were then exposed to photographic paper so that the developed film was darkest where the concentration of receptors was the highest. By means of a computer-assisted image-analysis system, the resulting pictures were color-coded so that the highest density of receptors appears white. Decreasing densities are red, yellow, green, blue, purple, and black. The high density of dopamine receptors in the substantia nigra of an 8-year-old boy (1) contrasts markedly with that in the brain of a 65-year-old man (2).

3-4. Receptor changes in disease
Such degenerative diseases of the brain as Parkinson's disease, Huntington's chorea, and Alzheimer's disease are characterized by a loss of specific neuronal populations. Associated with this loss of neurons is an alteration of many neurochemical features. The techniques described above have revealed some of these changes. For example, the substantia nigra in the healthy human brain contains high densities of the peptide neurotensin (3). In Parkinson's disease the loss of dopamine-containing neurons in the substantia nigra is accompanied by a marked loss of receptors for neurotensin (4).

the brain systems that control movement. The motor cortex and the thick bundles of nerve fibers that descend from it through the spinal cord serve principally to control fine movements of the fingers. Spherical masses of brain cells deep within the midbrain known as red nuclei (they actually are more pink than red in appearance) have the task of controlling limb movement. The organs of balance in the inner ear innervate the muscles involved in the control of posture through their own nerve-fiber pathways. Both cerebellum and basal ganglia are tied to all three of these motor subdivisions, for indeed, virtually all movement is a carefully orchestrated interplay of changes in limb position, finger grip, and posture.

Despite our extensive knowledge of the details of how nerve cells in the motor cortex, cerebellum, basal ganglia, and other parts of the motor system are connected to one another and to other parts of the brain, we still do not understand very well how they do such a fine job of coordinating the flexing and relaxing of muscles. We know that in Parkinson's disease, dopamine-containing neurons in parts of the basal ganglia degenerate. Yet we can do very little to help correct the patient's resulting inability to generate motor programs. Some of the most challenging problems of understanding the brain lie in its motor system. Overcoming them will require that we ask some very good questions indeed about how the brain manages our movements.

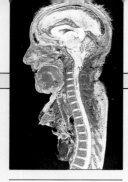

Brain and Motor Control

by Emilio Bizzi, chairman of the Department of Brain and Cognitive Sciences, Massachusetts Institute of Technology

Here I shall discuss experimental findings revealing some of the control solutions that have evolved for generating coordinated multijoint movements. Multijoint movements involve kinematic and dynamic issues.[1] The kinematic problem is that of trajectory. "Trajectory" refers to the path taken by the hand as it moves from one location in movement space to another and the speed with which the hand moves along that path. To plan an arm trajectory, the central nervous system (CNS) must solve a number of complex computational problems. For instance, in multijoint movements, the torque required to move one joint is dependent on the position of the other joints. In addition, there are joint-interaction effects that result from muscles spanning more than one joint. To further complicate the computational problem of executing an arm trajectory, the CNS must deal with torques resulting from interactional forces. As illustrated in figure A, if joint 1 is caused to rotate in the indicated direction, a reaction torque will act on joint 2. Similarly, joint 1 will rotate if a torque is produced at joint 2. Figure B illustrates also the centripetal interaction torque. Here segment 2 is assumed to have been caused to rotate about joint 2. The centripetal force acting on the mass establishes a reaction force (proportional to the square of the angular velocity at joint 2), which acts on joint 1 through segment L. The magnitudes of all these interactional forces are affected by the particular trajectory of the arm and can often be quite large.[2] If there is no compensation for these coupling effects, motion about one joint would cause other joints to flail, so that errors in joint motion and hand motion would occur.

These issues in multijoint control raise many questions. Is the trajectory plan specified in joint coordinates, or is the plan Cartesian, specifying the path of the hand and therefore requiring a transformation into joint coordinates? How are the necessary joint torques determined? Are they computed on a real-time basis, or is some memory approach used?

In an effort to gain some insight into the question of multijoint arm movements, neuroscientists have developed a variety of approaches ranging from physiological recording of single neurons in animals performing simple motor tasks, to kinematic studies of arm movements, to the study of how the CNS controls and takes advantage of the mechanical properties of muscles. The physiological recordings of single cells from central motor areas, an area of research developed by Edward Evarts about twenty years ago, have been invaluable in providing crucial information about the timing of single-cell discharges during a motor task. In addition, a number of interesting correlations have been established between neural activity and various parameters such as amplitude, velocity, and direction of movement.[3] While this technique has great value, it is clearly inadequate for understanding the computational problems related to the execution of movement.

Kinematic studies of arm movements have provided some information about planning. Recordings of planar arm trajectories have yielded two findings of interest when subjects move a hand from one target to another. First, the path taken by the hand is usually straight or only gently curved. This is of interst because the straight hand movement results from the combined effects of rotation of both joints; curved movements would result if the two degrees of freedom were not perfectly coordinated.

Schematic illustration of interaction torques

In A, a horizontal two-link arm is fixed at one end to the floor by way of joint 1 (J1). A second degree of freedom provided by joint 2 (J2). If J1 is caused to rotate in the indicated direction, a reaction torque will result in motion at J2. B illustrates motion at J1 caused by centripetal interaction torque (rf, reaction force; cf, centripetal force). (From Bizzi and Abend, in *Motor Control Mechanisms in Health and Disease,* ed. J. E. Desmedt, 1983)

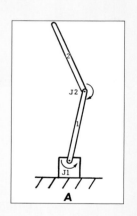

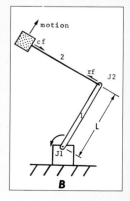

142

Therefore, the tendency to produce straight hand paths suggests that path planning must occur.

A second finding, reported by Morasso,[4] is that traces of joint position and joint velocity vary widely from movement to movement, while the speed of the hand is always roughly bell-shaped, even when the joint angular velocities are complex. The independence of the hand-speed profile from work space is consistent with the notion that the CNS plans a movement in terms of hand kinematics and then transforms the plan into joint coordinates.

The neurophysiological investigation of the processes underlying the transition from hand planning to the execution of a well-coordinated arm movement is an area of lively and exciting research. One promising strategy for attacking this problem involves the study of muscle geometry and mechanics. In fact, the investigation of muscle mechanics has been found useful in gaining insights into the rules of the neural controller. This approach is based on the assumption that the neural control system must not only control, but also take advantage of, the musculoskeletal apparatus.

A case for this appraoch was made years ago by Feldman, who investigated the springlike properties of the human arm.[5] Muscles do indeed behave like tunable springs in the sense that the force generated by them is a function of length and level of neural activation.[6] Several recent studies have emphasized this point and have stressed that we can understand the organization of voluntary movements by studying the way in which the mechanical properties of the neuromuscular system constrains the motor controller.[7]

Mussa-Ivaldi et al. recently provided a unique demonstration of the degree to which the neuromuscular system is springlike.[8] They have shown that when the hand is displaced from a posture and held briefly in a new position, arm muscles generate a restoring force that, when the arm is released, returns the hand to the initial position in a manner entirely compatible with what would be expected of a spring system. By recording the response to displacements of varying size and direction, they succeeded in describing the field of elastic force associated with the hand. At any given hand position, this field is characterized by a shape, a magnitude, and an orientation. They also found that the stiffness field associated with hand posture varies substantially at different positions of the hand in the work space.

These studies of multijoint arm posture and previous experimental work on visually triggered hand and arm posture in trained monkeys have shown that the position of the limb is an equilibrium point between sets of opposing forces.[9] This observation has led to the hypothesis that movements are centrally represented as gradual shifts in the equilibrium position of the limbs. The findings of Bizzi et al. suggest the existence of a gradually changing control signal during movement of the forearm from one equilibrium position to a final equilibrium point.[10] In the transition from the initial to the final position, the alpha motor neuron activity defines a series of equilibrium positions that constitute a trajectory whose endpoint is the desired final position.

The value and significance of these experiments lies in their suggestion that the CNS may plan movements in terms of the stiffness field of the endpoint of the linkage (the hand). The gradual transition between equilibrium positions would facilitate coordination of multijoint movement and reduce problems of control.

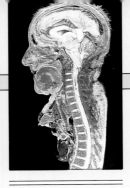

The Roles of the Cerebellum and Basal Ganglia in Motor Control

by Masao Ito, The University of Tokyo

Motor control of our bodies may be compared with modern technological control of such highly sophisticated vehicles as spacecraft. Hundreds of reflexes operate within a human body, much as numerous local controls of valves and flaps operate in a spacecraft to regulate levels, pressure, temperature, etc. The cerebral cortex generates central commands for voluntary motor control, as a pilot in a spacecraft issues commands to local control systems. However, with only these arrangements, our motor control would not work efficiently, just as a spacecraft pilot would be overloaded by numerous, complicated control tasks. And just as a spacecraft carries a large computer for solving this problem, so must our motor systems have a system to integrate the control of movement. The cerebellum and basal ganglia seem to provide this function.

The cerebellum accounts for about 10 percent of the total brain mass. Classical neurology and neurophysiology suggest that this organ assures smoothness and accuracy of movement through its ability to learn. Skills in performing the movements required in professional sports or musicianship require the aid of the cerebellum. The basic structures of neuronal circuits in the cerebellum suggest its learning capability. It is also thought that the cerebellum contains numerous subsystems, or "learning machines," each of which is connected to a reflex arc or a voluntary control system.

My colleagues and I have been studying the contribution of a cerebellar subsystem to a reflex that controls eye movement. The horizontal vestibulo-ocular reflex is driven by signals from the horizontal semicircular canals in the inner ears and produces eye movements that compensate for head movements. Our ability to read a book on a moving train is due to this reflex, as it acts to keep retinal images constant in spite of head movements. The horizontal vestibulo-ocular reflex arc connects with a small area of the cerebellum, the flocculus, which, on receiving semicircular-canal signals, modifies the reflex arc. Visual signals that represent errors in stabilizing retinal images are also sent to the flocculus and act to modify responsiveness of floccular neuronal circuits to vestibular signals. We hypothesize that the overall performance of the reflex is progressively improved by the error-triggered modifications of floccular circuits aimed at minimizing retinal errors.

Remarkable modification of the horizontal vestibulo-ocular reflex can be induced by wearing dove prism goggles, which reverse the right-left axis of the visual field, as demonstrated by Gonshor and Melvill Jones (see figure). In this situation the reflex is progressively depressed and even reversed in polarity. Animal experiments support this hypothesis by demonstrating that the adaptability of the vestibulo-ocular reflex is eliminated by lesioning the flocculus and that neuronal signals in the flocculus change along with modification of the vestibulo-ocular reflex.

The basic scheme of cerebellar control of reflexes that we derived from these experiments is that with its learning capability the cerebellum continuously calibrates reflexes to assure their accurate and smooth performance. Future investigations should reveal the exact manner in which a cerebellar subsystem assists voluntary motor control. A possible clue to this future comes from the recent, remarkable development of an active control technology for vehicles with on-board computers. In an airplane of this type, the pilot's commands enter a computer along with feedback signals from peripheral movement sensors. The computer then generates commands for optimal control of actuators. The major consequence of this new control scheme is that there are six degrees of freedom; there are normally four (thrust, pitch, yaw, and roll). The added two are direct side-to-side and up-down translocations. Since our limbs have many degrees of freedom (about 30 for a forelimb, including fingers), such control augmentation is essential for successful motor control. The basic idea that the cerebellum is essentially an organ for control augmentation seems to be a useful general concept of motor control.

The basal ganglia are assemblies of several neuronal masses. Hypokinesia in Parkinsonism and hyperkinesia in chorea, both arising from lesions of the basal ganglia, suggest that a major role of the basal ganglia is stabilization of motor systems; overstabilization may lead to hypokinesia, while loss of stabilization may produce hyperkinesia. Stabilization is a prerequisite for any complex, large-scale system, and stabilization augmentation is a major theme of modern control technology. Stabilization augmentation requires a strategy different from that for control augmentation. It is tempting to speculate that the basal ganglia provide stabilization augmentation in motor control, in contrast to the cerebellum, which is specialized for control augmentation. We lack sufficient knowledge of neuronal circuits and neuronal signals to immediately verify this suggestion, but its implications should help to guide future investigations into the still only vaguely defined functional purpose of the basal ganglia.

11
Brain Development: A Matter of Life and Death

How do the billions of cells in the brain form from the few primitive cells of the embryo and make proper connections with other cells? It all starts with a fertilized egg, just one cell. That cell divides to make two, they divide to form four, and the process continues until cells have specialized to form skin, bones, blood vessels, and the brain. At a very early stage some cells are designated to form the nervous system, a process known as neural induction, and begin to behave differently from the cells that will form the liver or the heart. Their genetic machinery is geared up to make chemicals for sending signals to other nerve cells and to put together the large, complex chemical antennas for receiving chemical messages from still other cells. Movement and growth of brain cells follow a complex choreography as cells glide toward their designated positions and form links to other members of the cast.

Surprisingly, the brain of a newborn contains far more connections than necessary. Brain cell death and the loss of connections are critical aspects of the formation of the brain. Since the adult brain contains something like a trillion nerve cells, and each of these cells is connected to about a thousand others, there is a real problem in assuring that the brain makes several quadrillion connections properly. Suppose that you had sent away for a computer kit that required you to solder together a quadrillion wires. Even if you lived long enough, could you get it right?

But suppose that the instructions for the kit told you to connect everything to everything else within reach. Then the device you had built would take care of the rest, removing the connections it doesn't need, leaving just the essential ones. But the start of the problem is even trickier. The wires of the brain, nerve fibers or axons, have to grow, and they must grow into the right neighborhood. Paul Letorneau of the University of Minnesota has given us some clues about what growing in the right neighborhood means. He coated plastic laboratory dishes with patches of two substances, one that growing axons stick to, and one that they do not. Onto this chemical checkerboard he placed very young neurons from a chick embryo and watched them grow. As they grew, the axons stretching out from these cells carefully traced a path that kept them on the sticky surface. From many experiments of this sort, we know that brain connections are made according to some sort of chemical road map. Researchers would like to know where to find a copy of this map.

In Saint Louis, Missouri, shortly after World War II, some of the clues to the chemical map of the brain first emerged. Rita Levi-Montalcini and Viktor Hamburger first realized that dying is part of growing for the brain. Cell death, we know today, is perfectly normal as the brain takes shape and gets rid of those neurons that do not contribute to the work it must do. In order to survive, Levi-Montalcini reasoned, the cells need some special nutritive substance, something called a trophic factor. She and her collegues discovered the first trophic factor, a substance known as nerve-growth factor. This substance is essential for the life of nerve cells of the peripheral sensory system and the sympathetic nervous system, which supplies many internal organs. Its effects on neurons from the brain and elsewhere in the central nervous system seem less dramatic but are still not clear.

The prevailing idea of brain development today is that there are many different kinds of trophic factors in the brain. Each trophic factor will maintain only a certain sort of nerve cell, so when the wrong sort of axon tries to connect with a cell, the trophic factor from that target cell will not nourish the neuron that gave rise to the errant axon. It is the cold-shoulder treatment in the extreme and leads to the death of the cell whose advances were rejected. One of the most intense searches in neuroscience today is the chemical treasure hunt for trophic factors in the

A carefree baby
The development of the brain in the embryo is one of the most fascinating puzzles that neuroscientists are attempting to solve. How does such an enormously complex organ arise from relatively sparse genetic information?

Brain Development: A Matter of Life and Death

The origin of the brain

Brain growth outstrips the growth of other organs. The cells of the nervous system are among the first designated to perform a special function early in the formation of the embryo. By the age of eight years, the human brain has attained its adult size. At the beginning of the fourth week of pregnancy (1), the human embryo, which is about 1.9 millimeters long, shows a neural fold in the ectodermal layer of cells that will form both the nervous system and the skin. When the embryo is 22 millimeters long (2), the neural fold has partially closed into the neural tube that will become the spinal cord and brain. By the time it has attained a length of 34 millimeters, the embryo begins to show more distinct features (3). In yellow are the neural tube, the cerebral protuberances, and the first nerve bundles. At the beginning of the second month of pregnancy (4), the embryo has attained a size of 42 millimeters. Noticeable at this stage are the prominence of the head, the growth of the cranial nerves, and the formation of the eyes. By the end of the second month (5), the different subdivisions of the brain and the eyes are clearly distinguishable.

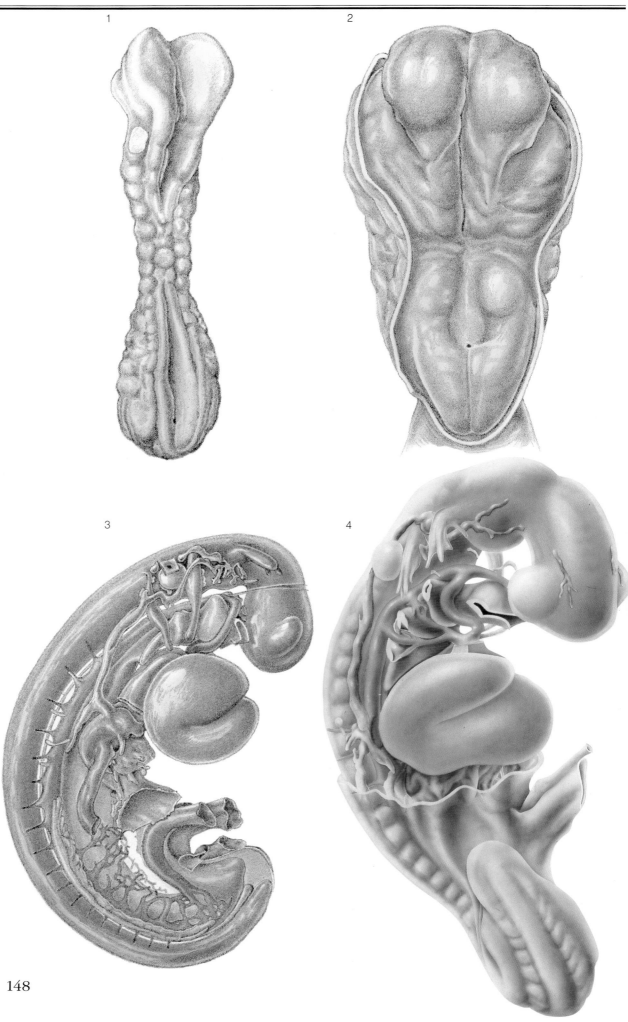

Brain Development: A Matter of Life and Death

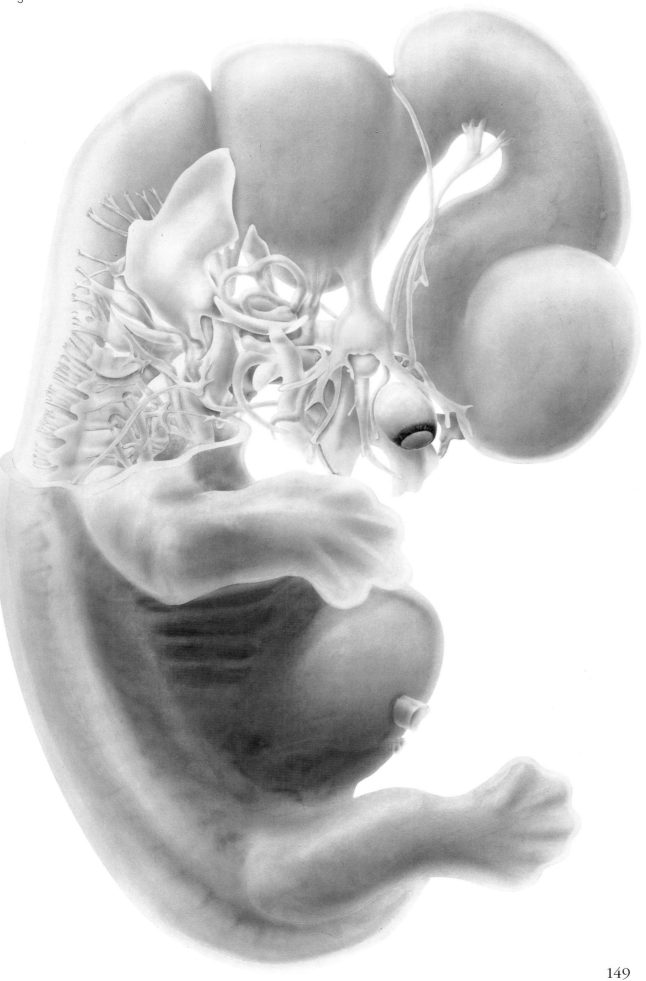

brain. The discovery of a genuine trophic factor may well lead to a Nobel prize for the scientist who tracks down the right molecule. The hunt, whatever molecules it yields, is quite likely to teach us once again that in trying to find out how the brain is organized, we have been asking the wrong questions.

Formation of the brain begins very early in life when cells of the embryo are just beginning to take on characteristics that distinguish one organ from another. The first definitive event, neural induction, takes place when cells from the outer embryo surface shaped like a hollow sphere begin to move to the inside, much like a soft balloon pressed by a finger. When the entering cells contact those remaining on the outer surface, some signal, presumably chemical, passes from the inner to the outer cells and activates specific genes in the cells that will become the nervous sytem. There is still much to happen.

Soon the flat sheet of designated nerve cells folds into a tube that will become the spinal cord. A bulge at one end of this tube will become the brain. The layer of cells lining this tube is where the action is at this stage. New cells are born from continuously dividing cells in this lining. The newcomers in the brain must then migrate in waves to the outer layer, which will eventually become the all-important cortex. Each successive wave of cells migrates further out than the one before it, so that cells in the outer layers of the cortex are younger than those deeper in the brain. Scientists now believe that long thin cells called radial glia that stretch from the inner to the outer surface of the growing brain act as guides for the migrating cells, but no one has yet discovered how the migrating cells know when they have arrived at their places of permanent residence. What seems likely is some combination of information programmed into the cell genes and information provided by local chemical signals.

Once in place, the brain cells send out their long fibers (axons) that will carry signals to other cells and the bushy structures (dendrites) that will receive signals from other cells. How the brain goes about wiring itself correctly, sending axons to proper cells, is a major puzzle in brain research. The death of cells, mentioned earlier, and the loss of inappropriate axons from cells that still maintain contact with other neurons appear to be largely fine-tuning processes; the brain could not possibly function properly unless its neurons knew fairly specifically where their axons were to go.

With so many things that can go wrong in the process of building a brain, it is not surprising at all that sometimes there is an error, usually with devastating consequences. While there are many known developmental disorders of the brain, including many different forms of cerebral palsy, one of the most recent to be pinned down is childhood autism. The afflicted child (usually a boy) is virtually cut off from the world. Previously thought to be unwilling to communicate, autistic patients prove really to be unable to function mentally. They may curl up in the corner, clasp their knees, and rock, or they may bang their heads against the wall. Often they are unable to use language and show insensitivity to pain along with heightened sensitivity to sound. Once thought to be a consequence of some sort of emotional stress, autism is now recognized as a biological disorder of brain development.

The first systematic study of the brains of persons who had sufferd from childhood autism was performed by Edward Ritvo of the University of California, Los Angeles. Compared to the brains of normal individuals, the brains of autistic patients show abnormal development of the cerebellum, with far fewer of the large cells knowns as Purkinje cells present. This finding is suggestive, because these cells serve, among other things, to modulate incoming sensations. It is only suggestive, however, because as we have seen, the functions of the cerebellum are as poorly understood as those of most other parts of the brain. Margaret Bauman of Massachusetts General Hospital also finds evidence in the brains of autistic patients that brain cells in the limbic system, which is involved in emotional functions, fail to grow apart in a normal manner. The story is not complete, but taken together, the evidence in so far indicates that the abnormal mental profile of the autistic child is the direct result of an error in the brain's developmental program, not of a flaw in the parents' treatment of the child.

Brain Development: A Matter of Life and Death

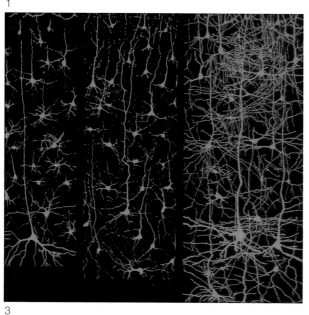

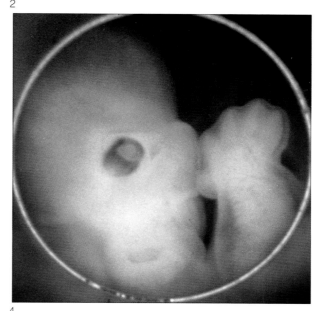

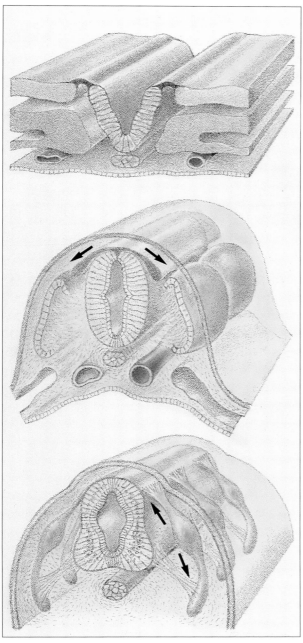

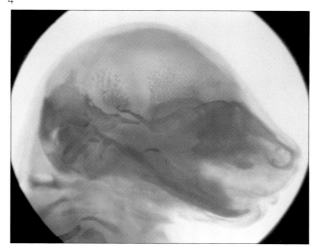

1. Connections
During the course of the first few years after birth, a network of complex connections forms among neurons in the brain. From left to right are pictures of the neurons in the cerebral cortex at birth, at three months, and at two years of age.

2. A living embryo of seven weeks
Ultrasonic imaging makes it possible to view the growth of the brain within the womb. The technique makes it possible to detect hydrocephalus and other disorders of development.

3. The formation of the neural tube
The three stages in the formation in the embryo of the neural tube (the early spinal cord) appear in this illustration. At the top the cells in blue, which have been designated to become neurons, form the neural plate, which begins to form the neural groove. The neural folds at the outer edge of the neural plate rise up and begin to bend toward one another. In the middle drawing the tube has closed, and the neural folds become the neural crests and move away from the tube. Finally, at the bottom the neural crests become the spinal ganglia, and these clusters of nerve cells send nerve fibers toward the periphery of the body and toward the spinal cord. These fibers will eventually carry sensory information from the body into the spinal cord, where it will be relayed to the brain. Failure of the neural tube to close leads to spina bifida and hydrocephalus.

4. The formation of the cranial vault
The considerable importance of the brain for the survival of vertebrate species has led to the development over the course of evolution of an effective means of protecting it. In this photograph of the head of a rat embryo three days before birth, one can see the formation of a delicate network of bony tissue in the membrane that covers the brain. Soon after the long bones of the skeleton have formed, the cranial vault will be complete, and the spaces between its bony plates (fontanels) will close. The fontanels are readily detected on the skull of newborn infants.

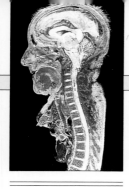

Cell Death in the Developing Nervous System

by Max Cowan, vice president, Howard Hughes Medical Institute

Although at first mention it may seem odd, it is now clear from a large body of work that in almost all parts of the nervous system there is an initial overproduction of neurons and a subsequent elimination of a substantial proportion of their numbers. It is now almost 80 years since the first published report of cells dying in the developing nervous system, but the significance of this phenomenon in the development of the central and peripheral nervous systems was not fully appreciated until the seminal work of Hamburger and Levi-Montalcini in the late 1940s. While studying the development of the dorsal root sensory ganglia in the chick, these workers found a consistent period in normal development during which close to 50 percent of the neurons in the sensory ganglia degenerated. Furthermore, if at a very early stage in development one limb was removed, virtually all the cells in the related sensory ganglia degenerated. They subsequently reported a similar, naturally occurring cell death in the spinal cord and suggested that this might be a rather general phenomenon in the formation of the nervous system. This conclusion has been amply borne out by subsequent work, which has established that in all but three neural structures that have been quantitatively analyzed, there is substantial cell death.

Neuronal death is now well documented in both invertebrates and vertebrates (including several structures in the human brain). In addition to being more or less ubiquitous, the phenomenon of naturally occurring neuronal death is one of considerable magnitude, amounting in some systems to as much as 85 percent of the initial population of cells. Moreover, the finding that the phase of cell death usually occurs over a rather restricted period during normal development, usually during the period when the cells are forming connections in the regions to which they send their axons, suggests that the establishment of new cell-cell interactions is an important factor in determining the fates of neurons. And the further finding that if the target field to which the cells project is removed at an early stage, there is additional cell death within the neuronal population indicates that the critical factor is to be found in the target regions rather than within the cells themselves. Taken together, these various observations have led to the hypothesis that in most parts of the nervous system there is an initial overproduction of neurons and that the axons of the neurons compete with each other for something available in only limited amounts at their sites of termination. One early idea was that the axons compete for specialized contact (or synaptic) sites within the target field, but later work has pointed to a more general explanation. This is that the axons compete for a diffusible growth or trophic factor available in only limited supply in the target area: those cells that are successful in the competition survive, whereas those that are unsuccessful die.

Direct evidence in support of this notion has come from the study of the only well-documented trophic factor, so-called nerve growth factor (NGF). NGF is a relatively small protein known to be essential for the survival of cultured sensory and sympathetic neurons. It has recently been shown that if exogenous NGF is administered to developing embryos, it prevents naturally occurring cell death in the sensory ganglia. For the rest of the nervous system, the relevant growth factors are not known. However, in the past three or four years several putative trophic factors have been identified, some of which have been quite convincingly shown to be able to maintain developing neurons grown in tissue culture. It seems likely that within a few years some of these factors will be shown to be equally effective for the intact nervous system.

One obvious biological role for naturally occurring cell death is that it matches the size of each developing neuronal population to the functional needs of the area to which it projects, whether another group of nerve cells, a sensory receptor field (like the retina of the eye or the sensory receptors in the skin), or an effector tissue like muscles or glands. In addition, neuronal death may also be an effective way to eliminate nerve cells that have made erroneous connections during their development. There is now good evidence that in almost every system some nerve cells may send their axons either along the wrong pathway, to the wrong target, or to an inappropriate region within the target field. In most cases virtually all of these aberrantly projecting nerve cells are eliminated during the phase of naturally occurring cell death.

In a few special situations, hormones seem to be critical for the survival of nerve cells. For example, most sexually dimorphic structures in the nervous system develop initially in the same way in both sexes; at some later stage cells become dependent for their survival on the availability of the appropriate sex hormone. If the hormone is not available during a critical period in development, there is secondary atrophy of neurons and most cells die. For the sake of completeness I should also mention that in many invertebrates cell death is strongly genetically dependent. In these organisms cell death serves to eliminate particular families of neurons, and in so doing, it regulates the total number of nerve cells. It is not known whether similar lineage-limiting cell deaths occur in the vertebrate nervous system, but there are reasons for thinking that this may be the case.

Finally, it is important to note that cell death is only one of the regressive events that occur during neural development. An important second class of such phenomena involves the selective elimination of particular axonal branches, while the cell that generated them remains alive. In most instances the terminal branches are removed, and this leads to a progressive refinement of the original pattern of connections, but in some cases even long collateral branches are eliminated. The mechanisms responsible for eliminating axon terminals and axon collaterals are largely unknown, but it may well turn out that they too involve a competition for trophic substances available in only limited amounts.

12
The Brain's Chemistry Set

The brain uses a great variety of chemicals to carry out its functions. Brain cells are not like liver cells, kidney cells, or lung cells. They are parts of an incredibly complex biological computer. This computer, however, is not electronic, at least not in the sense of today's science. The computer of the brain processes information not just in the form of electrical signals but also as chemical messages. Understanding the chemistry of the brain is therefore an essential part of learning how the brain works.

One important group of chemicals known as neurotransmitters carry signals between neurons. These affect the receiving cell by interacting with receptors. An electrical discharge flows down the axon of the sending cell and causes chemical changes that trigger the release of packets of neurotransmitter molecules into the synapse, a space that separates the axon from the receiving nerve cell. The molecules drift across this space only a few hundred wavelengths of light wide to arrive at the membrane surrounding the receiving cell. There the neurotransmitters mate with the receptors, and the resulting chemical events lead to electrical changes in the receiving cell. These changes may tend either to activate the receiving cell or to inhibit it. Activation may trigger a pulse of electrical energy to surge through the receiving cell, while inhibition will make the cell less likely to be activated by other signals.

This rather simple picture has not always been so obvious. In fact, during much of the first half of this century there were the most silly debates ever to occur in the history of brain research. Scientists, whose livelihood depends on the ability of other researchers to reproduce their results, should have known better. But instead of relying on the results of experiments, they taunted each other, not unlike children in a schoolyard.

A 1936 exchange between Nobel laureate John Eccles and W. Feldberg in Australia is revealing. At the time the experimental evidence favoring chemical transmission across synapses by means of acetylcholine was quite overwhelming, but many researchers, including Eccles, held to the view that electric impulses traveling across synapses activated the receiving cells. The difference was characterized as "soup versus sparks." The exchange proceeded in this fashion. Eccles to Feldberg: "Acetylcholine is all wet." Feldberg to Eccles: "Prefer wet acetylcholine to dry eddy currents."

Silly though it was, the debate marked a major turning point in the biology of the brain. Eccles went on to win a Nobel prize for his great work on the physiology of nerve cells and their "soupy" communications system. Yet almost a quarter of a century after this rather unscientific exchange of quips, the existence of true electric synapses was demonstrated.

Today we are no longer constrained to think in classical modes about communication among nerve cells. Just about any variation on the theme of which part of a nerve cell can send what kind of signals to which part of another cell has been demonstrated actually to occur in one part of the brain or another. Not only do axons send signals to dendrites or cell bodies, as Eccles's work so elegantly demonstrated, but dendrites can signal other dendrites and axons may be affected by signals coming from elsewhere on cells nearby. It is not an easy situation to understand, but the flexibility may actually make the ultimate task of describing how the brain works easier.

The principal neurotransmitters

As noted above, acetylcholine was among the earliest neurotransmitters to be studied. It occurs in cells in several regions of the brain, and reduction in the levels of acetylcholine is associated with Alzheimer's disease. Acetylcholine also acts as the transmitter between nerves and the skeletal muscles. The poison curare acts by blocking the receptors for acetylcholine on muscles; it paralyzes the victim

Brain slices
Here a neuroscientist traces the outline of a thin slice of a human brain. Similar slices from the brains of animals can be kept alive in special laboratory chambers for hours when supplied with oxygen and appropriate fluids. They display apparently normal electrical activity and respond to drugs and electrical stimulation, thus providing vital knowledge about the chemistry of the brain.

by blocking acetylcholine receptors. In fact, it was an ingredient from snake venom known as alpha-bungarotoxin that first made it possible for researchers to isolate acetylcholine receptors and determine how they work. This research also led to the discovery that the paralyzing disease myasthenia gravis involves an autoimmune reaction that leads to destruction of acetylcholine receptors on muscles.

In addition to acetylcholine, the "classical" neurotransmitters include a group of three known as catecholamines: dopamine, norepinephrine, and epinephrine (also known as adrenaline). These substances play a wide variety of roles in the brain and in various organs of the body and have functions related to emotion, attention, the control of movement, and the regulation of internal organs. There are several different types of receptors for each of the different catecholamines, a state of affairs that makes it especially complicated to understand how they work but has advantages for the development of drugs. The drugs known as beta-blockers, for example, selectively block epinephrine and norepinephrine receptors in the heart to lower blood pressure but have little effect on somewhat different receptors in other organs.

Dopamine is best known for its involvement in Parkinson's disease. Depletion of dopamine in the movement-control centers of the basal ganglia appears to be responsible for the disease, and most drugs used to treat it compensate in one way or another for this loss. The substance levodopa, for example, is used by the brain to produce dopamine and is the preferred treatment for most patients. The contrary condition, an excess of dopamine, is considered a possible cause of schizophrenia. And drugs like amphetamine that stimulate dopamine receptors can induce psychotic states in normal individuals.

The transmitter serotonin is similar to acetylcholine and the catecholamines in having widespread effects on many organs of the body. It was first identified for its ability to stimulate the contraction of blood vessels. Serotonin appears to be associated with many such basic functions essential to maintaining life as sleep, the control of body temperature, control of blood pressure, eating behavior, and drinking. The drug LSD (lysergide) appears to bring about hallucinations by blocking the action of serotonin.

Some very simple substances, amino acids, the units linked into chains to form proteins, also act as neurotransmitters. The best understood are gamma-aminobutyric acid (GABA), glycine, glutamate, and aspartate. Of these, GABA is particularly interesting, because it has a strong inhibitory influence on cells throughout the brain. Tranquilizers such as valium and other benzodiazepines appear to interact with GABA receptors to enhance this inhibitory effect. Huntington's disease sems to involve lowered levels of GABA in movement-control centers, which could be partly responsible for the loss of control that characterizes this hereditary disease.

Peptides, a group of chemicals that are formed by linking together as many as 100 amino acids, contain some members that act as neurotransmitters, but they also do many other things in the brain that we are only beginning to understand. Some of the earliest known peptides are produced by the brain to act as releasing factors to trigger the release of hormones by the pituitary gland. These include thyrotropin-releasing factor (TRF), corticotropin-releasing factor (CRF), and growth-hormone-releasing factor (GRF). These trigger the release of hormones that stimulate the thyroid gland (TRF), the outer layer, or cortex, of the adrenal glands (CRF), and tissues throughout the body (GRF).

Endorphins are particularly interesting neuropeptides. Their discovery resulted from the search to find where in the brain opiates like morphine exert their effects. Solomon Snyder and Candace Pert at the Johns Hopkins University, along with several other research teams, first found opiate receptors in the brain in 1973, which indicated that the brain must produce some form of opiate of its own. Such substances, discovered two years later by researchers in Scotland and Sweden, are now known to be distributed throughout the brain. The role of the endorphins in brain function is not yet well understood, but they seem to be involved in modulating pain, and they provide some of the most powerful pain killers known.

Knowledge of the chemistry of the brain has helped understand how some diseases result from chemical defects in the brain. It also holds out the hope that some incurable disorders like Parkinson's or Alzheimer's disease and others that involve degeneration of parts of the brain might be treated by transplanting new cells into the brain. It is well established that neurons in the brain and spinal cord of adults generally do not regrow once damaged. But research conducted during the course of the past decade, sometimes in the face of sharp skepticism, has begun to uncover some of the conditions that can indeed foster the restoration of lost neural function under laboratory conditions.

The current swell in activity in brain cell transplants dates back to 1971, when Joseph Altman, Gopal Das, and their associates at Purdue University demonstarted that embryonic neurons that had not yet begun to send out axons or dendrites could survive when transplanted into another brain. This early encouraging result seemed, however, to lead to a dead end. Mere survival of a neuron is not enough. It must also become a functional part of the information processing system that has broken down. The chances of this occurring seemed remote, for how could a small clump of cells transplanted into an adult brain re-form the precisely defined connections among other cells that the missing cells had made under developmental guidance?

The reality of some aspects of loss in the nervous system turns out to be not quite as grim as this inital assessment. Such rhythmic muscular activities as walking, for example, are now known to be largely under the control of interconnected groups of neurons in the spinal cord. That these pattern generators can function independently to generate seemingly complex activity can be demonstrated quite vividly in cats whose spinal cords have been severed. As expected, these animals are paralyzed and quite unable to walk. But if the animal receives an injection of L-dopa, the substance used to form the neurotransmitter dopamine, and is suspended with its feet on a moving treadmill, fully organized patterns of walking movements appear. So, at least in principle, a simple chemical input to a damaged neural system might restore its function.

Several experiments support the relatively simple idea of chemical replacement to treat the effects of neural degeneration. These studies mimicked the effects of Parkinson's disease in rats by destroying brain cells that produce dopamine on one side of the brain. As a result, the animals have a movement disorder that leads them to turn constantly toward the affected side. About a decade ago, a research team whose members came from

1. Brain enzymes
Progress in biochemistry makes it possible today for scientists to observe how enzymes direct the flow of chemical traffic inside brain cells. Without enzymes, reactions that now take seconds or minutes might take months or years. (Photo: Bristol Myers)

2-3. Seizure and trance
Brain activity involved in an epileptic seizure (2) and that involved in an hypnotic trance (3) may result in similar postures.

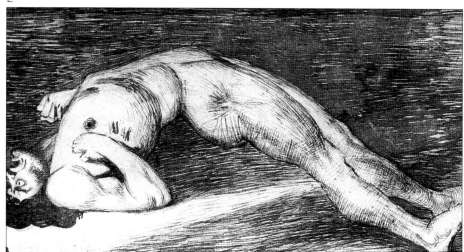

The Brain's Chemistry Set

1-4. The principal pathways of different neurotransmitters
In these side views of the median surface of the human brain, prepared by the central nervous system group of the Pierre Fabre Research Center in Castres, France, it is possible to follow the routes of nerve fibers containing the principal neurotransmitters. These routes make up (1) the noradrenergic system, (2) the dopaminergic system (the dotted region is situated more laterally), (3) the serotonergic system, and (4) the cholinergic system.

5. A computer model of a synthetic neurotransmitter
RS 86 is an agonist of muscarinic acetylcholine receptors. It is able to mimic the effects of the neurotransmitter acetylcholine in the brain and periphery. The figure is a computer-generated picture of the molecule of RS 86. Solid lines represent bonds between atoms; dotted surfaces represent the limits of the surface of the molecule. Different colors are used to represent different atoms. Red dots are the limits of oxygen atoms, blue dots are those of hydrogen atoms.

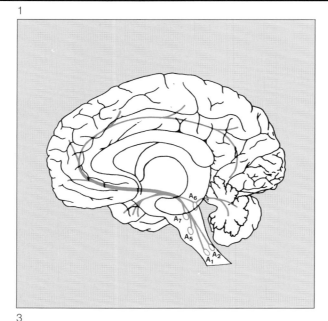
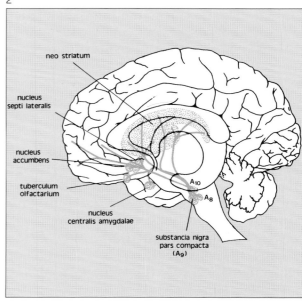
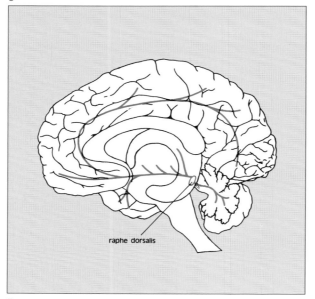
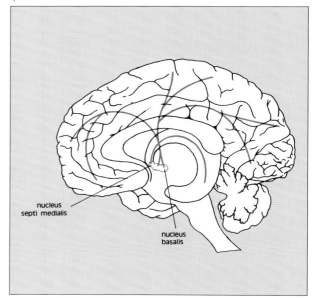
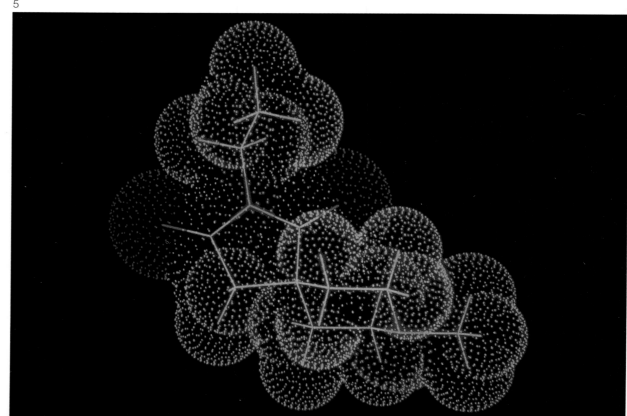

the U.S. National Institute of Mental Health (Mark J. Perlow, William J. Freed, and Richard J. Wyatt), the Karolinska Institute (Lars Olson and Ake Seiger), and the University of Colorado (Barry Hoffer) implanted in the brains of such animals dopamine-producing cells from the brain of fetal animals. The implants "cured" the circling behavior.

A more readily available source of transplantable material occurs in the adrenal glands. Cells in these walnut-sized glands perched atop the kidneys also produce dopamine. Freed and Wyatt inserted such cells in the brains of experimental animals to demonstrate that their insertion can also reduce the movement disorder. The results are not as dramatic, however, as those obtained with fetal brain cells.

After these and many other experiments, including several with monkeys, physicians concluded that they were justified in trying the technique on desperately ill humans with Parkinson's disease. The first human transplants were done by Olson and Seiger, working with Erik-Olaf Backlund at the Karolinska hospital. They transplanted cells from the adrenal glands into the brains of four patients. There appeared to be some slight improvement in the patients, who required less medication to reduce their movement disorders. Perhaps the most positive effect was the demonstration that the technique is feasible.

More recently, the level of excitement rose significantly as a result of a report from Ignacio Madrazo and his colleagues at La Raza Hospital in Mexico City. They too transplanted dopamine-producing cells from the adrenal glands of patients suffering from Parkinson's disease into the patients' brains, but they reported much more dramatic results. The difference between their results and those of the Swedish trial was thought to be due to the fact that the disease was not as advanced in the Mexican patients. Soon, however, the excitement began to fade as outside observers questioned the degree of improvement that had been reported. Still not discouraged, Madrazo continued to perform the transplants using tissue from the adrenal glands. He most recently reported further encouraging results using tissue from the brain and adrenal glands of a stillborn fetus. Though transplants using the patient's own adrenal glands have now become fairly widespread, a complete and balanced evaluation has still not been done, and the real value of the procedure remains unknown.

In the case of Alzheimer's disease, laboratory experiments with brain transplants have also produced some encouraging preliminary results. One of the principal characteristics of the brains of patients with Alzheimer's disease is the loss of neurons that contain the neurotransmitter acetylcholine in the hippocampus and in the cortex. Anders Björklund and his colleagues at the University of Lund cut the principal acetylcholine-bearing nerve fibers into the hippocampus of rats, impairing their ability to learn a maze. They subsequently transplanted either whole pieces of tissue or dissociated acetylcholine-producing cells into the hippocampus and showed that the transplants reduced the animals' learning difficulties. Similar experiments by Alan Fine of the Medical Research Council in Cambridge, England, Stephen B. Dunnett of the University of Cambridge, and Guy Toniolo of the University of Strasbourg showed that transplants could help reverse the memory deficits caused by removing acetylcholine inputs to the cortex.

Taken together, the results of these various experiments involving transplants into the brain suggest that the future will bring intensive research on the possibiltiy of using transplants to treat diseases of the brain. At present the most optimistic outlook for transplants into the human brain involves the use of fetal tissue. But effective therapeutic transplants are not likely to be achieved until many ethical and scientific issues have been resolved.

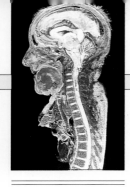

The Chemical Language of Interneuronal Communication

by Floyd Bloom, Research Insittute at Scripps Clinic, La Jolla, California

As one who comes to neuroscience from medicine, my natural objective in studying the molecules and molecular mechanisms that underlie signal processing in normal brains is to understand human brain disorders. I have created for myself three hierarchically related but separate concepts. At the most fundamental level is the concept of the structural organization of the brain in terms of the general sets or types of neurons and the molecules and mechanisms by which the interconnected neurons communicate. Above this molecular and cellular concept is a much less complete concept about the rules governing sets of interconnected neurons that regulate specific aspects of function. At a still higher and more incompletely formulated level of thought lie the rules, yet to be determined, that generate and regulate behavior and that, in humans, lead to conscious and unconscious mental activity. The construct at this level provides the framework within which mental disorders may become understandable.

The past two decades in neuroscience research have seen tremendous advances in the understanding of neuronal circuitry, in the identification of specific molecular messengers that distinct sets of neurons with certain morphological patterns of circuitry employ for transmisssion, and in the characterization of the molecular mechanism by which these chemicals act to regulate the target cells of the circuits. Together, the new circuits, transmitters, and receptor mechanisms provide a rich repertoire of signaling mechanisms that transcend earlier classical notions of excitatory or inhibitory transmission. I view the structural and chemical fundamentals as independent sets of three-unit categories. The most widely applied chemical classification of messengers divides them into three classes: amino acids, monoamines, and peptides. These classes present mole-to-unit-mass ratios roughly three orders of magnitude apart (micromoles, nanomoles, and picomoles per milligram of protein, respectively). While the molar-content ratios of amino acids and monoamines probably reflect the relative frequency of their sites of synaptic transmission, the low molar content of neuropeptides, lower by three orders of magnitude, is compensated for by the fact that peptides broadly coexist with many amino acid or monoamine transmitters.

Most neuronal circuits can be grouped into three general structural patterns: (1) neurons hierarchically arranged in chained systematic, or throughput, connections; (2) neurons diverging from a single cluster of homologous neurons as a source and forming connections with multiple target neurons; and (3) local-circuit neurons whose afferent and efferent connections are all located within a very small spatial domain (perhaps the most numerous single class of neurons). Those neurons that contain any given chemical class do not cluster on a simple spatial domain map as though they were functionally equivalent units of a single, coherent operational class. For example, some presumptive peptidergic cells are small interneurons, like enkephalin-containing cells, while others cover significantly broader spatial domains, like those cells in the CNS and peripheral nervous system containing vasopressin, somatostatin, substance P, or beta-endorphin. Thus, it would appear that spatial considerations per se do not correlate convincingly with either neurochemical diversity or the presumed functions of brain messenger molecules.

Monoamine and peptide transmitters produce novel actions unlike those of classically conceived transmitters, such as those of the neuromuscular juntion. These unconventional actions suggest that a broader range of action may be needed to examine transmitter actions. For example, when one examines in more complex experimental contexts the beta-adrenergic effects of locus ceruleus stimulation on other aspects of target-cell functions, the effects of the locus ceruleus no longer appear as a deviant form of inhibition. Rather, they appear to fit the description of "biasing" or "enabling" (the latter is a term I coined to indicate that the enabling transmitter [in this example, noradrenaline] can enhance or amplify the effectiveness of other transmitters received concurrently by target neurons).

These actions of norepinephrine would have been difficult to evaluate if the changes to norepinephrine had been examined alone. To recast this issue, I find it useful to designate both conditional and unconditional transmitter actions. Unconditional actions are those that a given transmitter evokes by itself (i.e., in the absence of other transmitters acting on the common target cell). Conditional actions include, but are not limited to, enabling. In such a conditional interaction each transmitter acts as its own postsynaptic transmitter receptor. They both act on the target cell when both transmitters occupy their receptors simultaneously.

It can thus be seen that there are abundant circuits and abundant transmitters. Moreover, many classes of chemically coupled systems exist to respond in different ways to the effects of active transmitter receptors. According to my concepts of molecular and cellular function, these receptors can operate either actively or passively, conditionally or unconditionally, over a wide range of time through nonspecific, dependent, or independent metabolic events. These and other possibilities reflect the fact that neurons have a broad and as yet incompletely characterized array of molecular responses that messenger molecules can elicit. The power of the chemical vocabulary of such commands lies in their combinatorial capacity to act conditionally, to act coordinatedly, and to integrate across the temporal and spatial domains within the nervous system. It is at this molecular and cellular level that many of the most exciting advances of our discipline have taken place.

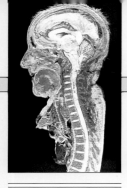

Hormones of the Brain

by Roger Guillemin,
Distinguished Scientist,
Whittier Institute for
Diabetes and Endocrinology, La Jolla, California

The word hormone is usually associated with the concept of secretion by endocrine glands. We talk of sex hormones, thyroid hormones, hormones of the pituitary gland, and also hormones of the gut (remember that historically the word hormone was coined for a substance present in the intestine and shown to activate secretion of gastric juice by the stomach). Hormones are molecules made in these various organs and carried by the bloodstream to other distant organs, where they exert profound functional control. Thyroid hormones, for example, control the metabolism of the overall organism, while sex hormones control the appearance of such secondary sexual characteristics as the development of muscles in males and the smoothness of skin in females. Much of the interest in hormones today concerns their functions in the brain.

The first brain hormones to be recognized were the peptide molecules made and secreted by nerve cells in the hypothalamus. Two of these hormones travel directly along the axons of nerve cells where they are made to the posterior pituitary gland, where they are stored until needed. One, antidiuretic hormone, regulates reabsorption of water by the kidneys; the other, oxytocin, causes contraction of smooth muscles to eject milk from the mammary glands, stimulates uterine contractions in females, and probably is involved in ejaculation in males.

Other hypothalamic hormones are involved in controlling the functions of the anterior pituitary gland, to which they are carried from the hypothalamus through a unique system of capillary vessels extending from the floor of the hypothalamus down into the pituitary tissue. The first of these molecules was isolated and characterized in 1968 after many years of research in several laboratories throughout the world. It is a peptide of three amino acids with the structure pyroglutamyl-histidyl-proline-amide and known as thyrotropin-releasing factor, or TRF. As its name implies, TRF controls the secretion of the pituitary hormone thyrotropin, which in turn controls the function of the thyroid gland.

We know now that for every pituitary hormone there is a controlling peptide of brain origin made in specific neurons of the hypothalamus and, as in the case of TRF, carried to the pituitary gland by the capillary portal vessels of the hypothalamo-hypophysial system. There is one such peptide that controls the secretion of adrenocorticotropin and of both gonadotropins (luteinizing hormone and follicle-stimulating hormone), and one that controls the secretion of growth hormone. As is often the case in physiological systems there is also an inhibitory peptide. This peptide is of hypothalamic origin, is known as somatostatin, and counteracts the peptide stimulating the secretion of growth hormone.

It came as a great surprise over the last few years to discover that these peptides generally recognized in the hypothalamus as specific controllers of pituitary functions are in fact widely distributed throughout the brain. They occur, not at random, but in highly specific groups of neurons that remain the same in the brains of all vertebrates, including humans. Extensive chemical evidence indicates that these molecules are indeed identical whether isolated from the hypothalamus, from the cortex, or other brain regions.

Even more surprising, one of these brain peptides, somatostatin, was found to be a powerful inhibitor of the secretion not only of pituitary growth hormone but also of the pancreatic hormones insulin and glucagon. Relatively simple calculations showed that it was impossible for somatostatin of brain origin to be involved in controlling the secretion of insulin, since the peptide somatostatin has a brief survival time in the circulatory system. The solution to the dilemma became clear when it was found that some cells of the endocrine pancreas also make and secrete somatostatin.

These cells are directly in contact with the cells of the pancreas, which are the sources of insulin and glucagon. This discovery was the first instance of a brain hormone found and characterized in a peripheral organ of the digestive tract. We know now that practically all peptide brain hormones are also found in such peripheral organs as the pancreas, the gut, the stomach, and in some cases the lungs and the skin. Conversely, the early-recognized peptide hormones from the gut have now all been found in specific neuron locations in the brains of all vertebrates, including humans.

Recently a series of peptide molecules have been found in various parts of the brain known to be involved in the physiological control of pain stimuli. Neurons in these regions are known to have receptors for opiates and to be responsible for the powerful analgesic effects of the opiates. The newly discovered peptide molecules are themselves endowed with all the pharmacological and biological activities of opiates. They make up a family found in the brains of all vertebrates and generally referred to as the endorphins. The exact mapping of all the neurons making one or another of these endogenous opioids is now well known. Here again, most of these molecules originally characterized in the brain have been found in peripheral nerves and also in such organs as the gut and the skin.

How do these brain hormones function in the brain? It has been known for some time that the conduction of nerve impulses from one nerve cell to another involves the release of simple molecules called neurotransmitters into the synaptic junctions between neurons. The concept of synaptic transmission developed entirely with the idea that only a few simple molecules like acetylcholine, dopamine, and norepinephrine act as the neurotransmitters. We now know that the various peptides discussed above are also to be found in neurons either alone or in combination with the classic neurotransmitters. There is increasing evidence that these brain peptides are used either as neurotransmitters on their own merits, so to speak, or that they are somehow involved in modulating the release of the classic neurotransmitters. It is still an open question whether the peptides in their multiplicity and ubiquity control and modulate the release of the classic neurotransmitters or whether these neurotransmitters may actually modulate the secretion of far more varied polypeptides at the level of the synapse.

One can heuristically view the highly variable peptides, obviously capable of carrying far more information than such invariant molecules as acetylcholine, to be involved in subtle higher functions of the brain. Such simple transmitters as acetylcholine may be only the keys to open the gates of the neurons, so to speak, allowing them to release or accept polypeptides from other nearby nerve cells. When one realizes that more than one hundred of these novel brain peptides have been chemically characterized over the last few years and that so little of their specific role in the function of the brain is currently known, one can only conclude that these brain hormones indeed pose one of the major challenges for neurophysiology and neuropsychiatry in the years to come.

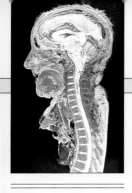

"Morphines" of the Brain

by Lars Terenius, Department of Pharmacology, University of Uppsala, Sweden

Morphinelike substances were recently discovered in the brains of mammals. This suggests that pain may be inhibited or at least modified by the body's own means. Under what conditions does this happen, and more generally, what does their existence tell about the nature of pain?

The subjective nature of pain

The purpose of a pain sensation is to protect us against (potential) tissue damage. Pain is therefore important: it warns and guides the individual. As a young infant develops, pain reactions have an essential role in learning. At all ages, pain raises negative emotions. It is the major reason why a person seeks a doctor.

In *The Puzzle of Pain* (1973), R. Melzack emphasizes the subjective nature of pain. A soldier wounded in war may feel relief to be sent back from the front and have no complaint of pain. During surgery, different individuals may require very different amounts of pain-killing drugs, so-called analgesics. The doctor can use only the patient's own descriptions and reactions to provide the patient with adequate pain relief. These and other observations show that pain is private and individual. To explain why pain sometimes appears less than expected, Melzack and Wall proposed that pain is controlled at certain points (so-called gates) of the pain pathways. The gate-control theory gives a conceptual underpinning to the subjective nature of pain.

A neurophysiologic model of pain

If some part of the body is injured, several kinds of nerves are activated. An immediate, sharp sensation travels in large-diameter nerve fibers and activates a reflex reaction. This is a warning pain. A second wave of pain travels in small-diameter fibers and arrives many seconds later. This pain is dull and diffuse. It may be called an ache and is the kind of pain that may become clinically unacceptable. Only the latter type of pain is sensitive to analgesics like morphine or aspirin. Severe damage to nerves conducting pain may lead to errors during nerve repair; a patient may then become oversensitive and experience pain after stimuli that normally would not cause pain. Such oversensitivity pain (neuralgia, causalgia) is frequently not sensitive to analgesics and tends to become chronic. Patients may even report pain in the absence of any observable tissue damage. There is clearly no simple relationship between stimulus and sensation (complaint). Pain may even be considered a mental state with clear existential significance and no neurophysiologic model is yet available to explain mentality.

A physiologic basis for pain modulation

Pain pathways have several nerve fibers in tandem; the synapses between the neurons are amenable to regulation via other nerves. The primary pain fibers terminate in the dorsal horn of the spinal cord (figure 1). The transmitters in these nerves excite the second-order nerves. Other nerves terminating in the same area could either affect the release of pain transmitter or directly affect the second-order neuron. Electron microscopy of the relevant synapses and electrophysiologic studies favor the latter model (figure 2). Several peptides (substance P, CGRP, somatostatin) are candidates as pain transmitters; they all excite the second neuron. Modulation occurs via endorphins (endogenous morphines) that are inhibitory. The interaction between pain transmitter and inhibitor becomes a simple matter of arithmetic. Endorphin nerve fibers are present throughout the neuroaxis and may affect transmission of pain also in various brain centers. Pain modulatory signals also travel from the brain to the spinal cord; some of these signals activate the spinal endorphin fibers. The systems described here constitute the postulated gates of pain control.

Endorphins are peptides and bear no structural resemblance to morphine, which is an alkaloid. Still, they act on the same receptors, albeit with different profiles. All three receptors modulate pain at a spinal level; in the brain the mu-receptor is the most important.

Peptides appear to be important both as transmitters for slow pain and for its modulation (endorphins). Peptides act more slowly and cause effects of much longer duration (minutes to hours) than nonpeptide transmitters. This may explain why pain characteristically builds up and fades away slowly.

Pain, stress, and placebo

Acute pain impels a defense response, which has obvious value for survival. Pain of some duration may be less purposeful and interfere with our ability to cope with the situation. Pain inevitably leads to stress. The stress reaction by itself is an adaptive response; interestingly, stress is also a trigger of pain suppression. In some kinds of stress, pain suppression occurs via endorphin release. However, stress may also reinforce pain by sustaining a pathologic process. A typical example is pain elicited by poor circulation in a swollen limb; stress worsens the condition by cutting the blood flow even further.

The pain reaction is highly dependent on emotional context. A trusting relationship between patient and doctor is very important; expectation of pain relief in a clinical setting is by itself a strong modifier of pain. This is an example of a placebo response ("placebo" is Latin for "I shall please"). For instance, the effects of a new analgesic drug are always controlled against a preparation identical in appearance, the placebo; the patient and usually the prescribing doctor do not know which is the placebo and which is the analgesic drug. The placebo response to procedures supposed to give pain relief is partly mediated via endorphins. It is, of course, respectable and in fact a doctor's duty to take advantage of the placebo response.

Acupuncture is a cornerstone of traditional medicine in the Far East, where it is used as an adjunct or the sole therapy of various disorders. A central concept in acupuncture is the correspondence between points on the body surface, which are stimulated by rotating needles, and inner organs. Energy is conveyed or withdrawn by the manipulating skills of the acupuncturists. In China, Mao Tse-tung requested the masses to study the richness of their old culture. Acupuncture was introduced for induction of analgesia. The Chinese experience was that acupuncture could be used as the sole analgesic in the operating room. Although less used today for surgical analgesia in China or elsewhere, acupuncture and related techniques, such as transcutaneous nerve stimulation (TNS), are established procedures for treating chronic pain. Such pain by definition resists other therapies. The results of acupuncture are particularly encouraging in patients who experience pain due to nerve damage. It is a common observation that one acupuncture session may be effective for a day or more. The techniques seem to operate at least partly through endorphin release. Like resetting the thermostat, acupuncture compensates a deficiency but does not cause an oveshoot, as pharmacotherapies tend to do.

Acupuncture and pain relief

We have seen that intense physical stimuli (like acupuncture) and physical or mental stress activate endorphin systems. Even expecting pain relief, a purely mental process, seems to be sufficient. What about the popular belief that endorphins are activated by jogging? There is experimental verification that endorphins are released during strong physical

Endorphin activation

exercise, like marathon running. There may be individual variations in pain sensitivity with age, sex, and cultural background. Ultimately, the endorphin systems are components in what we may call personality.

The pain pathway
The pain message travels through the pain neuron to the spinal cord; here it signals to a second neuron, which conveys the message to different brain centers. An enkephalin (ENK) nerve can affect the pain signal already in the spinal cord.

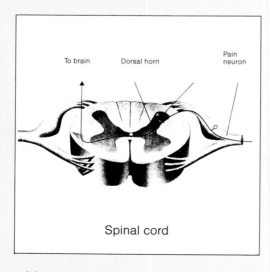

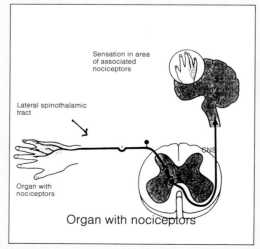

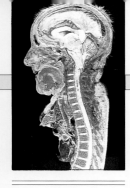

Brain Implants

by Anders Björklund,
University of Lund,
Sweden

Work on amphibians and fish was the first to demonstrate the possibilities for neuronal replacement after damage in the central nervous system. In such species, and above all in the visual system, it has been shown that grafted neurons can substitute both structurally and functionally for damaged neurons and that afferent and efferent connections can be established with a high degree of specificity between the grafted neurons and the host. In the 1940s Roger Sperry reported a series of classic experiments showing that vision can be restored in enucleated amphibians by transplanted or reimplanted whole eyeballs. These grafts reestablished a new pathway to the brain with correctly ordered connections.

For a long time it was thought that the remarkable regenerative and fucntional potential of CNS tissue grafts in cold-blooded vertebrates reflected a fundamental difference in the regenerative properties of central nervous tissue between cold-blooded vertebrates and mammals. During the last decade, however, it has become evident that at least certain types of intracerebral neural grafts can perform just as well in developing and adult mammals as in developing or adult submammalian vertebrates. This has led to the realization that, contrary to traditional views, the adult mammalian central nervous system has a potential to incorporate new neuronal elements into already established neuronal circuitry and that such implanted neurons can modify the function and behavior of the recipient.

Intracerebral grafting of monoaminergic and cholinergic neurons is particularly relevant to neurodegenerative diseases. In two neurodegenerative diseases, Parkinson's and Alzheimer's, the progressive and substantial losses of such types of neurons likely underlie some of the major clinical symptoms. Thus, in patients with Parkinson's disease the severe motor disturbances can be related to a loss of some 80 to 90 percent of the neurons in the nigrostriatal dopamine system. And in both Alzheimer's and Parkinson's diseases there is a strong correlation between the degree of dementia and the loss of cholinergic neurons in the basal-forebrain-projection systems, which are the sources for the cholinergic innervations of both the hippocampus and the neocortex.

Rats subjected to extensive experimental damage to the dopamine or acetylcholine systems are currently the most extensively used experimental animal models for Parkinson's and Alzheimer's diseases in man. In such rats, implants of fetal dopaminergic or cholinergic neurons have been found capable of forming new functional dopaminergic and cholinergic systems. This has been accompanied by significant, albeit incomplete, behavioral recovery in several types of motor, sensorimotor, and cognitive tests. Interestingly, this ameliorative effect has also been obtained with implants of dopamine- or acetylcholine-rich neurons in the striatum and hippocampus of behaviorally impaired, aged rats. The grafted cholinergic and dopaminergic neurons are capable of forming both electrophysiologically and ultrastructurally normal synaptic contacts with the initially denervated target neurons in the host, and it seems likely that the positive effects of the grafted neurons are due to their ability to replace the lost neurons and form new functional dopamine and acetylcholine pathways in the damaged host brains. More recently, comparable results with dopamine grafts have been obtained in monkeys with experimental Parkinsonism.

Grafting procedures

Good survival of CNS neurons has so far been obtained only from fetal or neonatal donors, while partial survival of peripheral ganglionic neurons is obtained also with adult donors. For CNS neurons, optimal donor age seems in general to be after the last cell division and neuronal migration but prior to the formation of extensive axonal connections. The donor age constraints for glial cells may be less critical, but this has so far not been well investigated.

Grafting is done either with small (one-to-several-cubic-millimeter) pieces of fetal tissue, which are implanted into ventricular spaces or into surgically prepared cavities, or with a dissociated cell suspension, which can be injected with a microsyringe by means of the so-called stereotaxic technique that allows precise placement.

In the mature brain or spinal cord, good survival of solid pieces of neural tissue depends on rapid vascularization from a vessel-rich surface (pia, choroid plexus, ependyma), exposure to circulating cerebrospinal fluid, and sufficient growth space. Neuronal-cell suspensions, prepared by mechanical dissociation of fetal neural tissue, seem to survive in any intraparenchymal brain or spinal-cord site. In the immature brain the conditions for graft survival and growth appear to be more favorable, and thus the constraints of graft placement, growth space, and graft size are less than in adult recipients.

Although the central nervous system is viewed as immunologically privileged, rejection of intracerebral grafts has been reported in grafts between species. On the other hand, good long-term survival of fetal CNS grafts can be obtained across major histoincompatibility barriers, such as between different mouse or rat strains (that is, under conditions when a skin graft, for example, is rapidly rejected). The partial immunological protection within the central nervous system is probalby attributable to several factors, such as the presence of the blood-brain barrier, the poor lymphatic drainage, and the low levels of major histocompatibility antigens expressed by CNS tissue.

Neural transplantation has demonstrated that the mammalian central nervous system has a remarkable capacity to incorporate and interact with implanted neuronal elements and that such implanted elements can modify the function and behavior of the host. This research has contributed significantly to the conceptual change that has taken place over the last decade with respect to the regenerative capacity of the mammalian central nervous system. It has thus gradually become apparent that under certain conditions CNS connections can be reestablished with high precision and over relatively long distances even in mammals. Thus, in sharp contrast to traditional views, the capacity of the mammalian central nervous system to repair and rebuild itself after damage may be considerable.

Neural grafting has so far not been successfully applied clinically to human cases of CNS damage. From experimental animal work, one is led to believe that one of the potentially most interesting clinical applications may be in patients with Parkinson's disease. So far only a few attempts have been made, with grafts of pieces of adrenal tissue taken from the patient's own adrenal medulla and implanted into the caudate or putamen of severely affected parkinsonian patients. Although some graft-induced effects on akinesia have been obtained, they were seen only acutely and did not last longer than two months. Thus, at this stage much research remains to be done to clarify to what extent intracerebral neural grafting may be a useful therapy in neurodegenerative diseases and in conditions of brain or spinal-cord damage.

Future perspectives

A potentially problematic issue is how to find an acceptable source of cells for neuronal implantation in man. At this stage, only cells derived from fetal CNS tissue have produced good results in experimental animal work. This raises the question of whether it would be ethically acceptable to use cells obtained from aborted, dead fetuses for transplantation purposes. In a longer perspective it may be possible to obtain cultured cells or cell lines of human or animal origin for transplantation purposes. Since clinical neurology has so little to offer in the way of effective treatment of various types of brain or spinal-cord damage, neural implants as a technique for neuronal replacement and CNS repair is a fascinating topic for current neuroscience research.

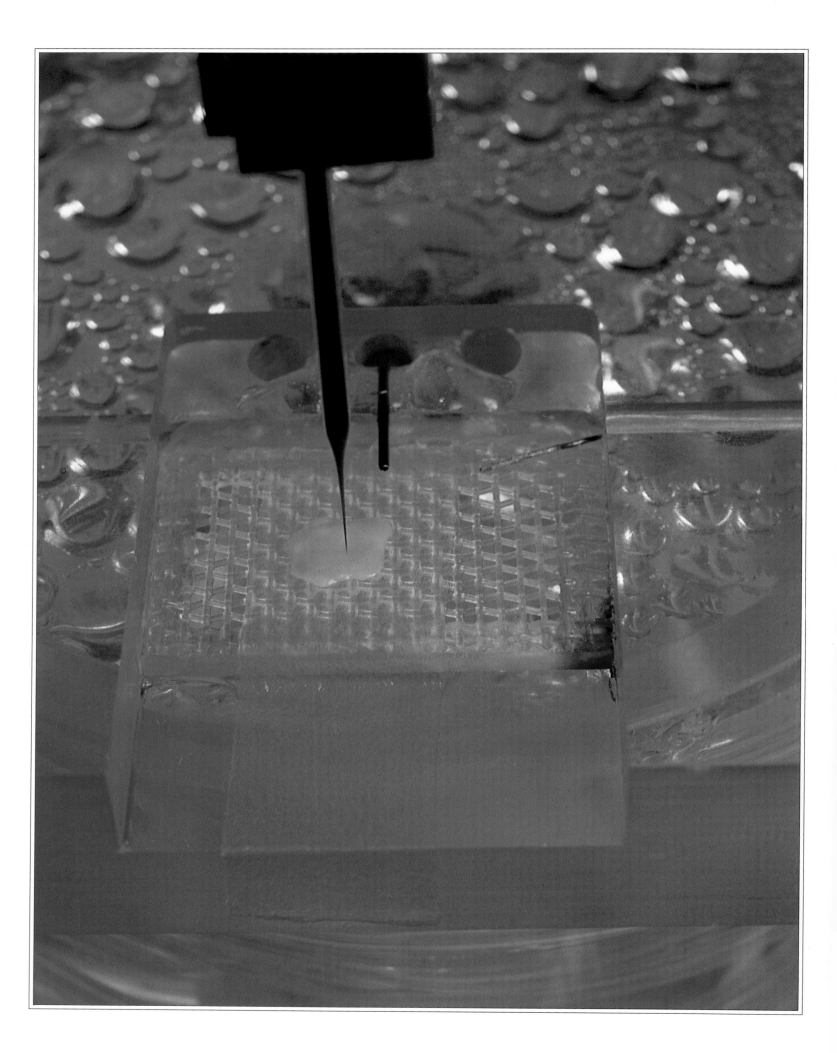

13
The Stuff Memories Are Made Of

The formation of memories is one of the most important of the brain's functions. The vast number of memories that a person can store is puzzling in a sesne, but consideration of the huge number of connections among brain cells leads to a number of possible ideas about using them as a memory bank. Research on simple animals is beginning to show how brains actually perform this task. Sutdies of brain regions involved in memory formation have begun to show us where memory forming and storing systems are located in the brains of higher animals. In memory research there is considerable excitement in the air, as well as optimism that perhaps we can do something at last to deter that most common complaint of aging, the dimming of memory.

Recent research shows that people have several different kinds of memory. One, the kind most people think of as memory, is known as declarative. It holds the names of things and people, the occurrences of events, and various other facts. We have some sort of direct contact with declarative memory and can talk about its contents. The other sort of memory, procedural memory, is a rather recent discovery. Because we are not directly aware of it and cannot talk about it, it was rather difficult to discover. People who suffer memory loss because of brain damage usually lose declarative memory. They cannot recognize new acquaintances, no matter how many times they meet them. The same joke is new each time they hear it. But they can learn to do things, to acquire new skills, like solving a puzzle, even though they don't know they have learned it. Knowing how to ride a bicycle involves procedural memory. If you have not ridden a bicycle in years, you probably cannot describe how to do it, but if you get on one, you'll find you remember how.

Anatomical and psychological studies of human memory are beginning to narrow down the location of the memory-storing apparatus. In a recent study conducted at the University of California, San Diego, and the Salk Institute, Stuart Zola-Morgan, Larry R. Squire, and David Amaral were able to study the brain of a patient known by his initials R.B. whose pattern of memory loss had been carefully documented over a five-year period. Following an interruption of the blood supply to his brain, R.B. became unable to form new memories. He could remember events from earlier in his life and could recognize famous faces. But he could not remember a new story for more than a few minutes after he heard it.

After R.B.'s death, researchers conducted a thorough microscopic study of his brain. Everything appeared quite normal except for clear damage restricted solely to the hippocampus. Evidence from the study of many other victims of brain damage and from animal studies has long implicated the hippocampus as being involved in memory formation. The case of R.B., however, was the first in which a clinically well-documented case of amnesia could be tied to a lesion restricted solely to the hippocampus, a significant step toward working out the circuits involved in memory formation in the human brain.

Where is procedural memory stored? University of Southern California psychologist Richard F. Thompson has recently found evidence that it is located back in the movement control center of the cerebellum. His experiments involve training rabbits in a standard conditioning routine. He sounds a musical tone and shortly thereafter releases a gentle puff of air into the rabbit's eye, a stimulus that causes it to blink. After pairing the tone with the puff for a number of trials, the animal learns to associate the two stimuli and soon will blink when the tone sounds without the puff. (Of course, if the tone continues to be a false alarm, the response goes away, or extinguishes, in phychological terminology.)

After the rabbit has learned to associate the tone with the puff, Thompson removes about one cubic millimeter of tissue from a deep region of the cerebellum known as the

Drugs to improve memory?
Scientists record electrical signals from neurons in a living slice of brain tissue while they apply new chemical agents they are testing for their possible beneficial effects on memory.

The Stuff Memories Are Made Of

1. Brain damage and memory
Lesions of the brain, like the one illustrated in this frontal section of a human brain, can wipe out the ability to learn.

2. Teaching a parrot
Conditioning techniques were used to train this macaw to shoot a toy cannon.

3. Bobo
In July 1982, Bobo, the oldest laboratory monkey in the world, made the front cover of Science. *Scientists at Lederle Laboratories used Bobo to study the effects of aging on memory. At 47 his age was equivalent to 120 human years.*

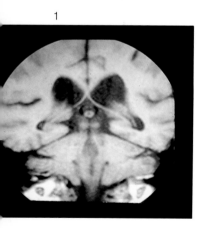

interpositus nucleus, and the memory is gone. The animal still blinks when the air puff occurs, so its movement control systems have not been affected. We still cannot tell whether the surgery has removed tissue where the memory is stored or just part of the neural circuitry invovled in expressing the memory. Applying to the interpositus nucleus kainic acid, a chemical that destroys cell bodies in a local brain region without damaging the nerve fibers that run through, has the same effect, which further supports the notion that the cells of that particular nucleus are essential. As this work progresses, it may prove that the cerebellum is significantly more versatile than was once believed.

The most fundamental aspect of what memory is must lie at the level of molecules and remains hidden within the innards of the genetic machinery of brain cells. We know from the work of Eric Kandel and others that chemical changes do occur in the nerve cells of an animal that has learned something. Some of the details of these changes are clear. Others are only speculation. Short-term memory involves chemical alterations of some existing molecules and the memory fades as these molecules slowly drain away. Long-term memory seems to involve changes in molecules that either make something new or make more (or less) of something already being made. This involves sending a special chemical message to the storehouse of blueprints for making chemicals, the bundles of DNA inside the nucleus, and gearing up the chemical machinery for the change in orders. We have a few hints about what might change in a brain cell as a consequence, and the increasingly powerful tools of molecular biology might reveal the secrets any day now.

One of the things that makes brain research so exciting is that it is possible to interpret new discoveries in many different ways. Because the storage of memoreis in humans and all other mammals remains such a mystery, what to some may seem a straight path to an answer may seem to others a misguided deviation. E. Roy John of New York University Medical Center remains one of the most vocal protesters to the idea that we can explain mammalian memory in terms of memory in sea snails and other simple creatures. Recent ingenious experiments designed by John and his colleagues show his point. They take advantage of the fact that a specially modified form of the sugar glucose, which is the principal fuel of brain cells, cannot leave the cells once it has entered. If a radioactive label is attached to the modified glucose molecules, experimenters can find out which brain regions are particularly active when an experimental animal performs a specific task. In these experiments, the task was a simple learning problem; cats learned to tell that food was behind a panel with concentric circles and not behind one with a star.

A particular feature of the experiments is that the two hemispheres of the cats' brains were surgically separated so that what the left eye saw was transmitted only to the right hemisphere, and what the right eye saw went only to the left. With this arrangement and the use of colored lenses, it was possible to conduct a two-step experiment. In the first step, one hemisphere could see the usual pair of star and circles on the two panels, while the other hemisphere saw only a meaningless triangle on each door. The cat was able to make the discrimiation quite normally, apparently taking advantage of only the memory stored in one hemisphere that the concentric circles mean food. During the second step, both eyes saw only meaningless triangles on the two doors, and the cat chose at random.

By using a different radioactive label during the two steps, John was able to compare the distribution of brain energy demands under two different conditions. During the first, one hemisphere is retrieving a memory about performance while the other is subject to nonspecific influences. By subtracting the differences, it is then possible to show what energy demands are associated with performance of the learned task. The results indicate than an estimated 5 million to 100 million neurons increase their level of activity in association with the learned performance and these are widely distributed throughout the

the brain. John concludes from this result that the standard notion of memory being associated with changes in specific neurons is not acceptable. Rather, he favors some as yet unrecognized cooperation among extremely large numbers of cells, and proposes that memory is a property of the brain as a whole, not of its elements.

What the mechanism of this cooperation might be is not at all clear. An idea favored by physiologist W. Ross Adey of Jerry L. Pettis Memorial Veterans Hospital in Loma Linda, California, is a form of communication among brain cells that differs from the standard picture of bursts of energy and spurts of chemicals used to carry messages. Adey's proposal is that more subtle electromagnetic fields are amplified by complex physical processes at the surface of brain cells. These electromagnetic "whispers" then alter the internal chemical machinery in some way that changes its role in processing information. Some evidence for such subtle effects comes from Adey's studies, which reveal chemical changes in the brain as the result of exposure to weak microwave signals.

Many researchers contest the views of brain function proposed by John and by Adey, citing lack of evidence and flawed techniques. The debate is a lively one, and the sheer complexity of the brain is likely to keep it raging for years to come.

4. Training for special duty
This Labrador was specially trained for the American army by Animal Behavior Enterprises. His task is to detect land mines buried less than 45 centimeters below the surface. He must then sit down and await the arrival of military personnel, who will defuse the mine. The dog's weight is distributed in such a way that it does not set off the mines.

5. Dance of the rooster
A dwarf rooster "dances" against a backdrop to tunes from a "jukebox." He first pulls a cord on the jukebox that starts his musical accompaniment and then begins to dance. It is actually a conditioned response that involves the delivery of an earthworm when the cord is pulled. The "dance" is nothing more than the animal's naturally occurring behavior of scratching for food, which is reinforced by bits of food delivered during the performance.

6. Aquatic ballet
A killer whale and a porpoise were trained to perform together at the former Marine World in California. Normally, killer whales hunt porpoises.

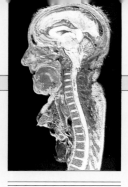

The Brain: A Self-Organizing Learning System

by Wolf Singer, director of the Max Planck Institute for Brain Research

The encephalization that accompanies the phylogenetic emergence of mammals led to a dramatic increase in the number of nerve cells and neural connections. The human brain contains somewhere between 10^{12} and 10^{14} nerve cells. If these were to form a chain connected by their cellular processes, this chain could be wrapped around the earth's equator about 20,000 times. The neuronal cables that establish connections between individual cells attain a total length of several kilometers within a cubic millimeter of cerebral cortex. These numbers exemplify the combinatory complexity of possible relations that can be established by neuronal connections and raise the challenging question of how such a complex system can develop in a safe and reproducible way from a comparatively small set of genetic instructions.

The development of organs and organisms is based on a permanent, interactive exchange between the information stored in the genome and the information available in the cellular environment of the genome. The latter determines through biochemical signals which of the genetic commands should be expressed. As development proceeds, the microenvironment of the genome changes, and this in turn leds to modifications of gene expression and to further changes of the microenvironment, and so on. While this self-organizing process is similar in all organs, the brain has the unique ability to use electrical activity as an additional signaling system in the developmental dialog between genes and their environment. Mechanisms are implemented that enable these electrical signals to influence gene expression. This has a number of extremely important biological and ultimately even philosophical implications. Electrical signals are transported by neuronal processes over great distances and with high topological specificity. Furthermore, from a certain developmental stage onward, the brain possesses functioning sense organs that convert signals from within the organism and even from extracorporal space into electrical messages. This enlarges dramatically the range and the complexity of the environment available to the self-organization process. The environment relevant for brain self-organization ultimately includes all the domains with which the evolving brain is capable of interacting. Another important aspect is that the very same electrical signals that convey sensory messages are used by the brain as information carriers for computational processes. Hence, the unique capacities of nerve nets to perform complex logic operations on large data sets also become available to the self-organization process. As the complexity of the brain increases, both its interactions with the environment and its computational capabilities become increasingly sophisticated. As a consequence, the set of parameters influencing further brain development becomes more complex and capable of supporting self-organization in conformity with even more differentiated structures. Because of this spiral of reciprocal interactions between the genome and its increasingly complex environment, a rather small set of genetic rules suffices to promote the devleopment of systems as complex as the human brain.

Learning to see

Most of our knowledge of experience-dependent developmental processes comes from the visual system. Clinical evidence and deprivation experiments indicate that visual centers develop normally only when visual experience is available during a critical period of early development. Patients who have suffered from congenital opacities of the eyes during early childhood and are therefore unable to perceive contours fail to recover visual functions after puberty when the optical media of their eyes are restored by surgery. The reason is that the genes alone are not capable of instructing the development of neuronal connections with sufficient precision.

Higher mammals and man have the ability to fuse the images in the two eyes into one percept and to compute from the differences between the images the distance of objects in space. This important function requires an extremely high degree of precision in the neuronal connections that link the eyes to nerve cells in the cerebral cortex. The one million nerve fibers arriving from each eye have to be arranged such that only those pairs of afferents that originate from corresponding loci in the two eyes converge onto common target cells in the cerebral cortex. Nature solves this selection problem by exploiting the fact that nerve fibers originating from corresponding loci in the two eyes are usually stimulated by the same contours and hence convey activity patterns that are more similar than those transmitted by fibers originating from noncorresponding sites. The required precision in connectivity is achieved by an experience-dependent selection process that stabilizes those afferents from the two eyes that convey correlated activity and disconnects those that do not. Such activity-dependent selection is a very powerful and versatile developmental mechanism for establishing enduring relations between nerve elements that share particular functional properties. In our special case, however, additional requirements have to be fulfilled for such selection to be successful.

Experience-dependent pruning of retinal connections may occur only when the animal is looking with both eyes onto a nonambiguous target, and it must not take place when the two eyes are moving in an uncoordinated way. In this latter case, the images processed by the two eyes are different and hence all retinal signals, even those originating form corresponding retinal loci, are uncorrelated. The consequence of pruning would be a complete disruption of binocular connections. The selection process must therefore be gated by nonretinal control systems capable of determining the instances in which retinal activity may induce changes in circuitry. There is again good evidence from animal experimentation suggesting that the mammalian brain possesses such gating systems.

Vision-dependent modifications of neuronal connections do not solely depend on retinal signals but require in addition the presence of gating signals that are generated by systems controlling arousal and attention. When these systems are inactive, visual signals fail to induce circuit modificaitons. Conversely, direct electrical activation of the arousing systems greatly facilitates vision-dependent modifications of neuronal transmission in the cerebral cortex.

The modulatory pathways responsible for this gating of neuronal plasticity originate from structures in the central core of the brain that are phylogenetically much older than the cerebral cortex and simultaneously influence large areas of the brain through widely distributed axonal arbors. The input to these systems comes from a large number of different brain centers. The "print now" command that enables circuit modifications can thus be influenced by the functional state of many other, nonvisual structures.

On the basis of these new insights into experience-dependent self-organization processes, the classical nature-nurture question of inherited versus acquired abilities can be readdressed, and some preliminary

Nature versus nurture

answers can be formulated. In the visual system and most likely also in other subsystems of the brain, the genome determines the general layout of neuronal connectivity and the basic response properties of neurons, many of which develop independently of experience. These experience-independent specifications predetermine the criteria according to which the brain selects and categorizes sensory signals. Since patterns of sensory activation have to match these prespecified response properties to induce long-term modifications, the genetic predisposition limits the range of stimuli that can influence self-organization. The genes also determine the nature of possible relations that can be established by experience between the various feature domains. Associations can be established only between those sets of neurons that are interconnected to begin with. Furthermore, the genome defines the rules according to which predetermined connections are selectively stabilized. We have seen that these modification rules have an associative function: they reinforce connections whenever two events occur simultaneously in space and time. Coherency, then, is one of the fundamental principles according to which the building blocks of the brain become interconnected. Likewise, coherency in space and time is also the criterion according to which associations are established among the phenomena of the outer world. Through the selective stabilization of coherently activated neuronal connections the brain builds internal representations of such relations. Objects and, in a more general sense, virtually all spatial or temporal patterns are distinguishable from the background only because they possess some coherent properties. The unique ability of the nervous system to evaluate such coherencies is therefore an extremely well-adapted function.

We have seen that experience-dependent modificaitons of connectivity depend to a critical extent also on internally generated gating signals. Thus, large neuronal arrays participate in the decision whether a particular activation pattern should lead to long-lasting modifications of circuitry. Experience-dependent self-organization, therefore, has little in common with passive instruction of a *tabula rasa*. Rather, the developing brain appears as a highly active and primarily self-contained system that, at birth, already possesses substantial knowledge about the structure of the world to which it is going to have to adapt itself. As mentioned above, this knowledge is stored in the brain's architecture and in the rules allowing for activity-dependent modification of this architecture. After birth when the brain is confronted with a dramatic expansion of accessible environment, it poses a number of precise questions to this enviornment to optimize and adapt its internal structure to reality. In a number of neuronal systems these questions are raised only during a brief and critical period. If answers are not available, the prospective functions do not develop, and these deficits are in most cases irreversible. Thus, there is an early and very active dialog between the developing brain and its enviornment. It soon becomes impossible to distinguish between cause and effect of certain developmental bifurcations. The developing brain and its environment appear as components of a highly interactive system whose complexity and richness exceeds by far the information provided by the genome. Because of this cooperation between the genes and their environment, structures as complex as the brain can evolve and acquire such sophisticated function. In this context it may be appropriate to raise the philosophical question of whether consciousness and mind can be considered as a property emerging from individual brains alone or whether brains of similar complexity must interact with one another during a long period of ontogenetic self-orga-

Self-organization and learning: The brain as an ever-changing organ

nization until mutual reflections between brains lead to the emergence of symbolic representations in the form of language and eventually to such concepts as those of the self and consciousness.

Another fascinating aspect of experience-dependent self-organization is its formal similarity with learning. At a descriptive and perhaps even mechanistic level the activity-dependent refinement of connectivity patterns shares characteristic features of associative learning. In the case of vision, two sets of variables, the afferents from the two eyes, become permanently associated with a common effector, the cortical target cell. The criterion for this association is coherency of activation. Interestingly, the same pattern appears in classical, or operant, conditioning, in which associations are established between sensory signals and behavioral responses. A rabbit responds with a blink when an air puff is directed at its eyes. If the air puff is frequently presented together with a buzzer tone, the tone alone will eventually lead to the blink. The tone has become associated with the air puff and acquires predictive value. Evidence is available that the newly formed associations in this and related conditioning experiments result from selective strengthening of neuronal connections that are frequently activated together. The concurrent activation of acoustic and somatosensory afferents appears to lead to selective strengthening of preexisting acoustic pathways until they alone are capable of activating the common effector cells that command the blink response.

Another similarity between developmental self-organization and adult learning is that sensory signals have to match some predispositions of the learning organism to establish associations. In the development of vision the minimal requirement is that the retinal signals conform with the response properties of cortical neurons. The same is true in adult learning. Stimulus configurations have to fulfill some minimal criteria for saliency and disambiguity in order to be effective. Finally, there is the fascinating parallel that selective associations occur only when the stimulus patterns co-occur with internally generated gating signals that are available only when the brain is awake and attentive. As we all know from introspection and as has been established in numerous behavioral experiments, new relations are learned only if the brain is motivated to pay attention to the respective stimuli. Even the chemical signals involved in this attention-dependent gating of associative functions appear to be similar in developmental self-organization and adult learning. There are indications that activation of the very same central core structures that facilitate connectivity changes in the visual cortex during development also enhance associative learning.

Developmental plasticity differs in two respects form adult learning. First, there is a critical period for the expression of the former. Second, in the developing brain, but not in the adult brain, connections are removed if they lose in activity-dependent competition. Yet there is one special form of learning, imprinting, that is also restricted to a critical period of early development. The best explored examples of this type of learning have been studied in birds. Young birds learn the songs of their parents, memorize the patterns of their plumage, and later select their mating partners accordingly. This learning process is confined to a critical period of early development and is very resistant to later modification. The adaptive value of temporal windows for self-organization processes is obvious, at least for vision. Once binocular correspondence is established, this variable must become fixed to serve as a frame of reference for high-level processes that use differences in retinal correspondences for the computation of spatial depth. One of the reasons for a temporal window in imprinting may be that it reduces confusion by confining the learning phase to a period during which the probability is high that only members of the family will be around.

In view of these numerous similarities between experience-dependent self-organization and learning, it is not always possible to distinguish the two processes. Clearly, imprinting can be interpreted in both ways. Maybe one should therefore consider adult learning as a continuation of some of the self-organizing processes that underlie the development of the nervous sytem and its adaptation to the environment. Certainly, in the adult brain some of the developmental processes are no longer active; the formation of new connections and the physical removal of established connections occur only in a few special cases. Yet existing connections continue to be modifiable in their efficacy in a use-dependent way, and the rules by which these modifications occur appear to be similar to those in developmental self-organization. It thus appears that the brain is an ever-changing system.

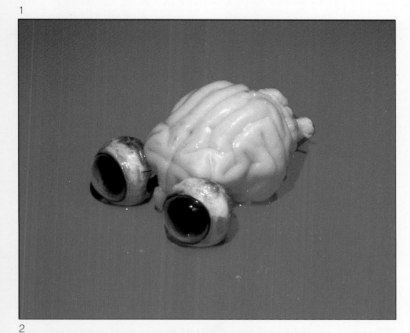

1. Optical connections
A cat's brain that has been prepared with the optic nerve connections between eyes and brain preserved.

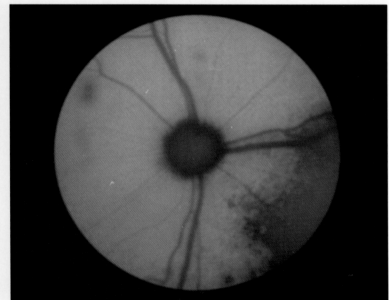

2. The retina
The circle in the center of the retina is the exit point of the optic nerve, which contains about one million fibers, and of blood vessels that nourish the retina. This area is not sensitive to light and accounts for a blind spot in visual space. We normally do not notice the blind spot because the brain receives information about the same location from the other retina. To detect the blind spot, close one eye and hold a pencil up so that its point is in the middle of your field of view. While looking straight ahead, slowly move the pencil toward the side of the open eye and note that the point disappears when it is slightly off center.

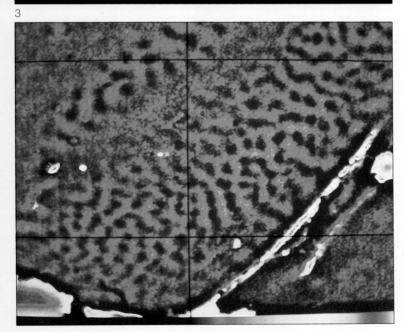

3. Neurons responding to vertical contours
To visualize this zebralike pattern, those areas in the occipital part of a cat's brain that contain nerve cells responding selectively to vertically oriented contours have been labeled with radioactive glucose. The cerebral cortex was subsequently removed, flattened out, and exposed to X-ray film. Since radioactivity darkens the film just as light, the pattern on the film reflects precisely the distribution of radioactivity in the brain. The resulting black and white distributions were subsequently processed by computer for quantitative evaluation. For such regular patterns to develop, the animal has to have normal visual experience with contours during early development.

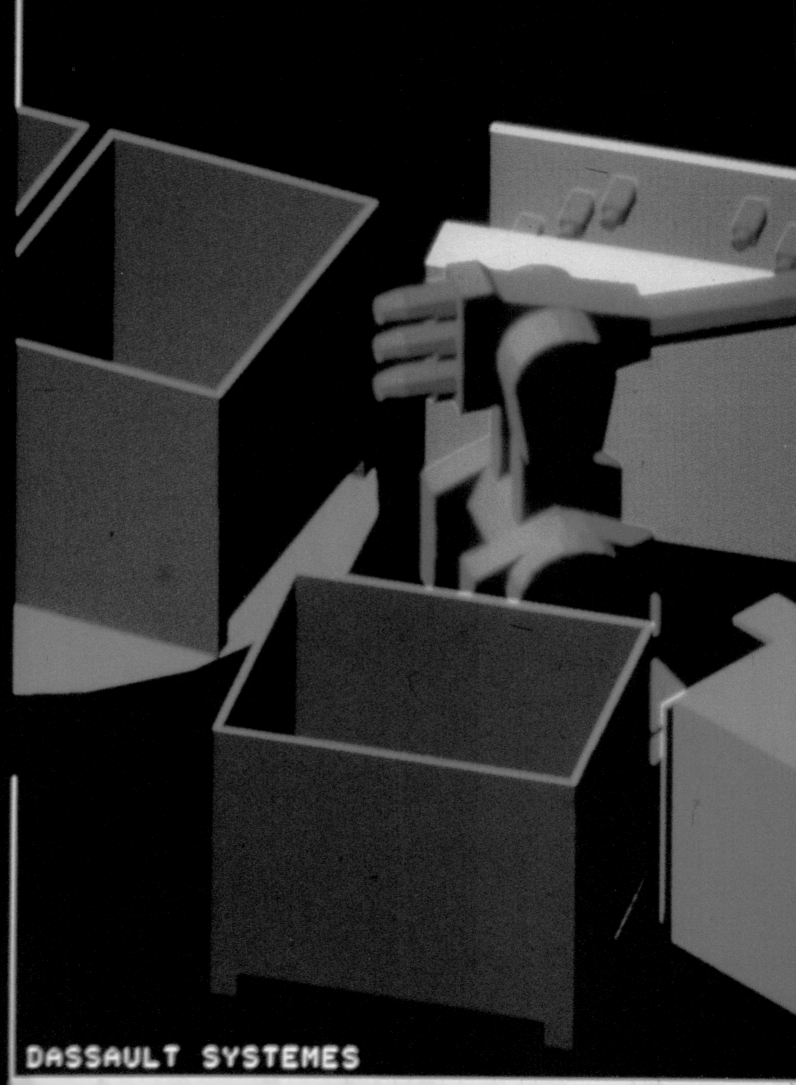

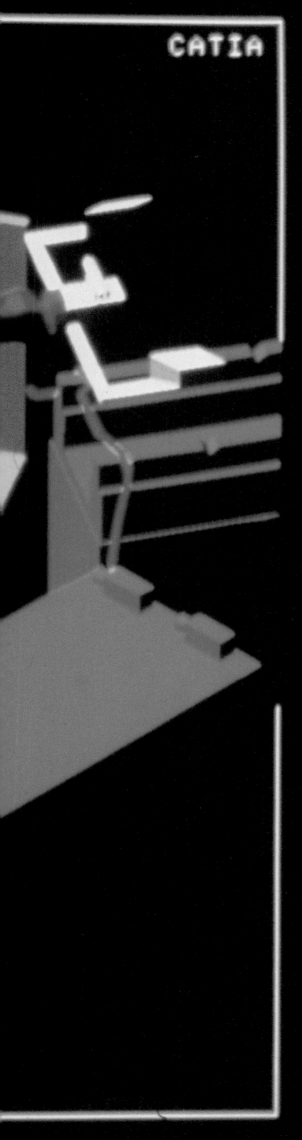

III
MIND-BRAIN CONNECTIONS

At this stage of our study of the brain, we can try to put back together the pieces we have been taking apart and studying one by one. Part 1 looked at very difficult, ambitious questions about consciousness and intelligence and suggested how to ask questions that might have good scientific answers. Then in part 2 we began to take the brain apart and carefully to examine some of its pieces. From this exercise we were able to learn something about how the brain works and gained a sense of what kinds of answers we might expect from asking proper questions about the machinery of the brain. Now we can return to some of the concerns that we raised in our early considerations of how brains and minds might be related. In this section I shall examine some of the most exciting current research revealing glimpses of what is really going on inside our heads.

A robot and its environment
Computer simulations of how robots behave help scientists to design actual machines and view their behavior in three dimensions. Illustrated here is an example of a problem caused by a collision of the robot with its surroundings.

14
Thinking Healthy Thoughts: Are You What You Think?

The idea that the mind can influence the body is not a new one. History is rich with anecdotes about how a healthy mental attitude can conquer disease. An ancient Egyptian papyrus considers the subject, and popular books on the subject become best-sellers today, an indication of just how eager we are to believe in the power of our minds. But finding good hard evidence of how the brain interacts with the biological functions of the body has not been an easy task.

The lure of hypnosis

Some of the most striking examples of apparent domination of the body by the mind come from the puzzling field of hypnosis. Hypnosis has had the lure of the occult about it from its very beginnings. The behaviors that appear to occur under hypnosis faintly suggest access to some otherwise unreachable corners of the mind. Can such behavior teach us something about the nature of the mind? Does hypnosis access remote corners of the mind? And does hypnosis have anything to tell us about how the brain works?

Hypnosis as we know it today dates back to the work of Franz Anton Mesmer in the eighteenth century. He published what apparently was a largely plagiarized medical thesis entitled *On the Influence of the Planets* in 1766. Early in his medical practice in Vienna, he found that he could effect cures by passing magnets in the vicinity of the patients' bodies. Soon he found that his waving alone was effective, and he dispensed with the use of magnets. He then moved to Paris, where his treatments became so popular that he introduced the practice of group therapy. But Mesmer failed to enlist the support of the medical community of Paris, and in 1784 a royal commission appointed by Louis XVI found that there was no detectible physical agent involved in Mesmer's cures.

From that time on, hypnotism followed a tortuous course. Medical practitioners employed hypnotism to anesthetize patients for surgery. Magicians used it on stage to entertain the masses. Freud at first embraced it in his therapy and later abandoned it as being too unreliable. Sometimes it worked, sometimes it didn't, and sometimes after it worked, it wore off. Then after Freud, hypnotism sank into scientific shadows, though it still enjoyed great popularity on the nightclub stage.

Scientific interest in hypnosis began to revive after World War II. In the late 1950s and early 1960s American psychologists like Martin Orne of the University of Pennsylvania and Ernest Hilgard of Stanford University attempted to apply the methods of experimental psychology to hypnosis. Their work led to the current view of hypnosis as a behavioral-social interaction that leads to alterations in perception, memory, and voluntary actions. Their work also confirmed that there is much variability among individuals in the susceptibility to being hypnotized and that we must be extraordinarily careful in interpreting the results of any experiment involving hypnosis. Once again, it is a confirmation of the enormous biological variation in characteristics of individual human beings.

While we may conclude that at least for the present there is very little of scientific value in hypnosis, we cannot ignore its parctical value. Throughout its history there has been a common thread in its application to anesthetize people. Practitioners of the art who followed Mesmer performed surgery with the aid of hypnotism, and the evidence suggests that the procedure relieved unthinkable suffering and saved many lives, despite criticism of the the authorities at the time.

The discovery of ether and other general anesthetics displaced hypnotism but did not entirely eliminate it from clinical practice. Physicians generally agree that hypnotism is unreliable for surgical anesthesia, but sometimes there is no alternative. Don Morris and some of his collegues at Louisiana State University recently reported a case of an obese woman with a huge benign tumor on her leg.

Growing brain cells in the laboratory
To develop methods to alleviate chemical imbalances in the brain, researchers grow isolated brain cells in laboratory dishes. These cultured cells are important tools for the development of drugs that can help dysfunctions of the brain.

Thinking Healthy: Are You What You Think?

1

2

3

4

Thinking Healthy: Are You What You Think?

1-4. The young fiancée of Villejuif
These four drawings by a manic-depressive patient dramatically illustrate the course of his illness. He expresses calmness (1), excitation (2), depression (3), and aggression (4). The materials and methods vary: ink (diluted or not), pencil, pen, and brush; sometimes there is shading, and sometimes not. The lines may be smooth or sharp, curved or angular.

5-6. Carl Frederick Hill
This Swedish painter completely changes his technique and his subjects during states of psychic dissociation. During such states he expresses delirious ideas and hallucinations, symbolic expressions of separation and restriction.

Her generally unhealthy condition made her an unsuitable candidate for general anesthesia, so the doctors decided to try hypnotism. The procedure proved quite successful. Only when it was necessary to attach a hydraulic lift to the 25 kilogram tumor did she show any signs of discomfort.

Just as sleep and general anesthesia by chemical means remain beyond the explanations of current brain science, so too do the psychological phenomena associated with hypnotism. At the moment it is unlikely that the study of hypnotic phenomena will shed much light on how the brain works. But any satisfactory explanation of how the brain works must take account of these phenomena.

Psychoneuroimmunology

One of the most complex links between mind and body that scientists are exploring today is how the body's main line of defense agginst disease, the immune system, might be affected by the brain. Immunologists once tended to scoff at the idea, but some are beginning to change their minds in the face of accumulating evidence. Marvin Stein of the Mount Sinai School of Medicine in New York City, for example, finds that lymphocytes are affected in recently bereaved widowers. For several months after the loss of their wives there was a decline in the ability of their lymphocytes to fight off infection. Further studies with laboratory rats revealed that if they are given inescapable electric shocks, their lymphocyte count drops and so does the ability of their lymphocytes to replicate, an essential part of the immune response.

Other experiments implicate the brain more directly. Robert Ader of the University of Rochester School of Medicine and Dentistry has taught rats to lower their own immune defenses. He uses methods of classical conditioning, presenting cyclophosphamide, a powerful immune system suppressant, in association with saccharin. After a number of training sessions, presentation of the saccharin alone evokes the suppression of the immune system just as if the cyclophosphamide had been given.

Just how the brain might affect the immune system remains an unsolved puzzle, but hormones from the pituitary gland are clear candidates for links. Also, the lymph glands, the thymus gland, bone marrow, and other organs that affect cells of the immune system all receive nerve fibers whose functions are not well understood.

The reverse link too is a problem. How does the immune system act on the brain? Hugo Besedovsky of the Swiss Research Institute in Davos-Platz, Switzerland, finds that hormones released by lymphocytes cause a reduction in levels of the neruotransmitter noradrenaline in the hypothalamus of the brain. Karen Bulloch of the University of California, San Diego has shown that nerve fibers from the thymus gland run to clusters of nerve cells, the nodosal ganglia, that cary information from internal organs into the brain. So there is clear evidence that the brain and the immune system can communicate with each other, a communication link that is clearly an important one in maintaining health. There may well be many ways in which we can make better use of this link to improve human health in the future.

Some novel ideas about this communication system come from J. Edwin Blalock of the Unviersity of Alabama, Birmingham. He proposes that cells of the immune system and cells of the brain both produce similar or identical hormones and have receptors for these hormones. The suggested hormones include corticotropin, the substance originally identified as carrying a message from the pituitary gland to the adrenal glands in stressful situations, and thyrotropin, the hormone from the pituitary gland that activates the thyroid gland and other glands.

Blalock's proposal is that the cells of the immune system serve as a special sensory system. When they detect microorganisms or viruses that might cause disease or the cells of tumors (which, of course, are not perceived by any known parts of the nervous system), they release one or more of these hormones. These hormones then convey an appropriate mesage to cells in the brain that can respond in some way to its content. Of course, this is only a suggestion at the moment, but in view of what has been learned so far, it is not a far-fetched one. Other novel ideas are sure to emerge rapidly as more and more researchers take up the challenge of exploring the effects exerterd by these once-hidden links between health and the brain.

Thinking Healthy: Are You What You Think?

1. Mental illness in the Middle Ages
The Extraction of the Stone of Madness

2. Mental illness today
The computer makes it possible today to detect, store, and compare the electrical activity of the brains of mentally ill patients. The results are used in the development of expert systems that can aid the clinician in diagnosis and treatment.

Thinking Healthy: Are You What You Think?

1. The zodiacal man
Watercolor and ink drawing from a Persian manuscript of the tenth century. For medieval man each sign of the zodiac was associated with a part of the human body. The sign of the ram, for example, goes with the head. In such a system of belief there would be no coincidence in the fates of a number of famous people born under this sign: Lenin died of a stroke; Henry II of France was killed by a lance in the eye; Baudelaire suffered paralysis and aphasia from a stroke; Van Gogh became insane; and Jayne Mansfield was decapitated in an auto accident.

2-3. Altered perceptions
The world perceived by the mentally ill undergoes striking alterations, as suggested by this sculpture and this drawing.

4. Queen of the animals
A major milestone in the study of the drawings of the mentally ill was the book Bildnerei der Geistekranken *(Imagery of the Mentally Ill), published in 1922 by German psychiatrist Hans Prinzhorn. He proposed that it is possible to learn much of the nature of mental illness from the form and the content of pictures such as this drawing by a psychiatric patient. The combination of humans and animals with sexual overtones is a common theme.*

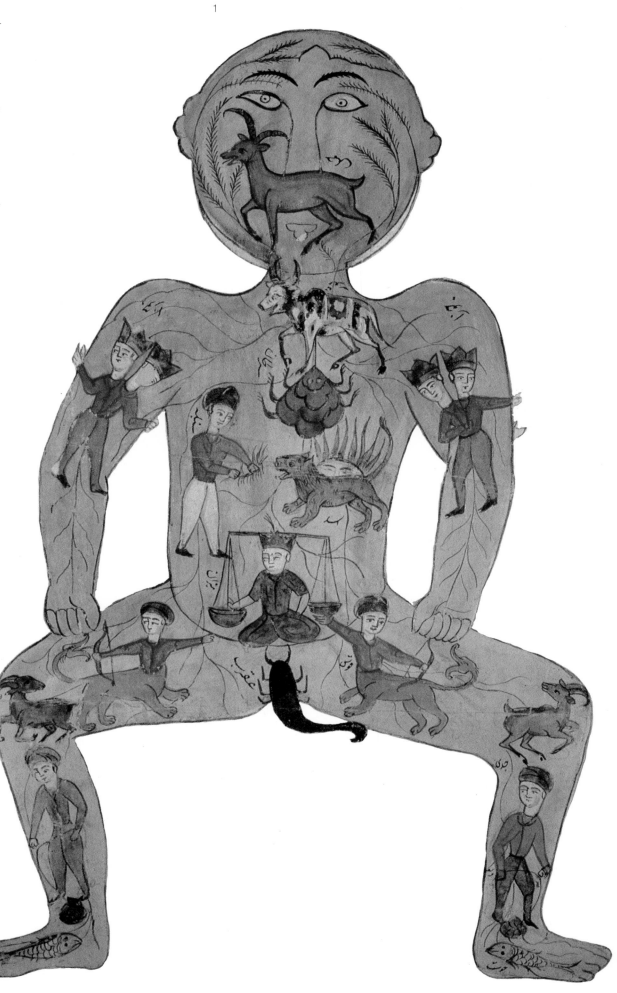

1

Thinking Healthy: Are You What You Think?

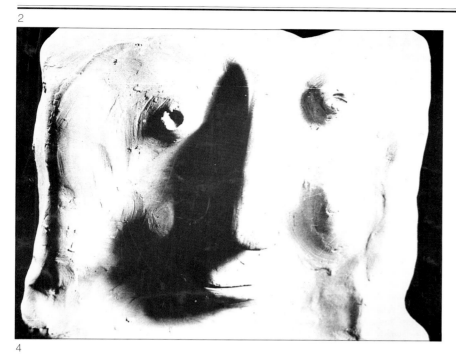

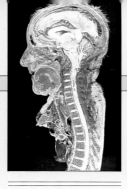

Paradoxes of Depression

by Pierre Pichot, Hôpital Sainte-Anne

Probably no other field of psychiatry has been so much researched over the last thirty years as that of depressive states. One reason for this is their frequency: at any one time some 6 percent of the general adult population suffers from depression. This figure has led Gerald Klerman, for example, to describe the modern era as the "age of depression." Clinical research has modified our understanding of the boundaries of this syndrome and the different forms it takes; effective biological and psychological treatments have been discovered; and biochemistry may well have furnished us with a fairly exact model of its cerebral and biological mechanisms. It does seem paradoxical that despite the amount of work done and the high quality of the results, there are as many questions to answer about depression as there were before. In other words, the solution seems to recede as fast as answers are found.

Depression is a syndrome whose symptoms may be divided into two groups: a feeling of sadness and pessimism, and a general slowing down of vital processes. Sadness is often associated with feelings of guilt and suicidal impulses. The general slowing down may be manifesed in a number of different ways, including a lack of enjoyment and interest, fatigue, difficulty in carrying out one's usual tasks, a modification of motility (generally, a decrease in motility), a feeling that one's intellectual capacities are impaired, appetite disorders (generally, a loss of appetite), loss of weight, and sleep disorders. In addition, the patient suffers from anxiety, often in an intense form.

Depression may be either symptomatic or primitive. Symptomatic depression is a consequence of some other illness or disorder (such as a brain tumor or viral infection) or one expression of some other form of mental illness, like schizophrenia. In primitive depression the depressive syndrome alone constitutes the illness. There are three major forms: the depressive personality (a permanent but relatively mild condition), endogenous depression, and neurotic depression. Both endogenous and neurotic depression generally take the form of short, intermittent episodes lasting from a few weeks to a few months; they account for the majority of primitive depressions.

Endogenous depression, or manic-depression, is generally intense with very marked symptoms. As its name suggests, it tends to occur for no apparent reason. Between episodes the subject behaves normally, and his past history will include either one or several similar episodes (the unipolar form) or an alternation of depressive and manic episodes (the bipolar form). This disorder would appear to have a biological cause, with heredity playing an important role, and it should react well to biological therapies. Neurotic depression is less intense, and the symptoms differ. It is generlly triggerd by some sort of trauma, which may or may not have serious effects (therapists talk about reactional depression). It appears to affect individuals with a fragile personality and neurotic traits, hence its name. The cause and nature of this type of depression are above all psychological, with no specific hereditary cause. The most suitable form of treatment is psychotherapy.

Today, however, this simple analysis is called into question. Researchers question in particular the absolute distinction between endogenous and neurotic depression, whch is the basis of the traditional view underlying most recent biological work. American research tends to show that there is no difference in the importance of hereditary factors in the two forms, and in the ongoing controversy characterizing the last twenty years or so of reasearch, particularly in Great Britain, the conclusion is that there may be only one type of depression and that its endogenous or neurotic aspects correspond to the extreme points of a continuum in which the majority of cases are located somewhere around the middle.

Clinical theory and practice as built up since the turn of the century is thus under fire, and the situation is no less uncertain in the field of biology. The basic foundations were provided by the completely empirical discovery of effective therapeutic procedures: electroshock therapy in 1938, the first tricyclic antidepressants (impramine, in 1957) and monoamine-oxidase inhibitors (iproniazide, in 1957). The classic hypothesis of biogenic amines to describe the mechanisms of depresion was developed in the 1960s by establishing a relationship between the therapeutic effects of drugs in man and their pharmacological activity in the animal brain. According to this theory, depression is linked to a lowered level of amines. In some synapses tricyclic drugs block the removal of biogenic amines (noradrenaline and serotonin), whereas monoamine-oxidase inhibitors act to prevent these amines from being destroyed. Unfortunately, this simple and attractive theory has run into a few snags. Pharmacological reaction in animals is very fast, whereas therapeutic effect in the human subject can be seen only after 15 to 20 days. A drop in the rate of cerebral amines can only be demonstratd indirectly (by testing urine or cerebrospinal fluid, for example), and there is considerable room for error. It is probable that many other neurotransmitters, some of which remain to be discovered, play a role in synaptic transmission. Furthermore, although neuropharmacologists have developed many antidepressants with a specific action (for example, drugs that block uptake of only serotonin), clinical tests have not shown that their antidepressive effects differ markedly from those of other drugs, either quantitatively or qualitatively. The large amount of biological research done on this subject for over 30 years has resulted in definite experimental results, but these have often merely rendered the situation even more confused. Most researchers still consider that depression is characterized, as far as the brain is concerned, by a disturbance in synaptic transmission, but the nature of this disturbance is contested (a hypothesis enjoying current popularity implicates hypersensitivity of postsynaptic receptors). Moreover, although researchers have discovered a number of new antidepressants, some belonging to different chemical families, their advantages over existing drugs may reside simply in the fact that they have fewer side effects, since none of them can be proved to work better than impramine and amitriptyline, which were discovered at the end of the 1950s. Finally, despite the many hypotheses advanced, no one has yet been able to discover any explanation for the relative success of electroshock therapy.

The whole field of depression is therefore a paradoxical one. Empirical discoveries have transformed therapeutic procedures, but the clinical and biological research they have stimulated, while advancing the state of our knowledge about depression, has at every stage revealed that our ideas are in need of revision. If we are to make any progress in this field, researchers must be fully aware of all these areas of uncertainty.

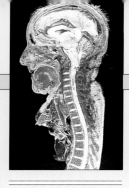

Is the Brain Involved in the Elaboration of Immune Responses?

by Kathleen Bizière, Hoffmann-La Roche and Co., and Gérard Renoux, University of Tours

Most people believe that they are most susceptible to illness if they are sad or worried. This belief implies that the central nervous system (CNS) is involved in mounting an immune response. Surprisingly, what appears to be a commonsense statement is based on very little experimental evidence. Thus, few studies in humans have examined whether emotional factors can affect the immune response, although there have been reports showing that stress and bereavement impair immunity.[1] The same is true in animals, and until recently the question of whether the CNS could modify immune responses had practically not been addressed. We shall briefly review animal studies that clearly suggest mutual interrelationships between the CNS and the immune system.

Endocrine modulation of immune responses

Hypophysectomy has been shown to cause involution of the thymus, a gland vital to the immune system, and to alter the activity of thymus-derived cells of the immune system.[2] Conversely, thymectomy has been shown to affect pituitary functions. In addition, there are numerous reports indicating that hormones modify immune responses. It is thus generally admitted that the endocrine system can modulate the functioning of the immune system. Since the endocrine system is controlled by the CNS, it seems probable that the CNS is involved in immune responses. The question remains whether the CNS also controls the immune system directly or whether the CNS acts only indirectly through the endocrine system.

Neural modulation of immune responses

In keeping with the observation that hypophysectomy can alter thymus activities, it has repeatedly been shown in animals that lesions of the anterior hypothalamus impair the activity of antibody-secreting B cells.[3] Lesions of the posterior hypothalamus, on the contrary, depress the activities of T cells, which organize immune response and carry immunological memory.[4]

The sympathetic nervous system (which is under the control of the CNS) also seems to be implicated in mounting an immune response. Such lymphoid organs as the spleen, the thymus, and the lymph nodes receive rich sympathetic innervation. Destruction of the sympathetic nervous system alters immune responses.[5] In the spleen and thymus, sympathetic nerve endings have been shown to form an intimate, synapselike relationship with lymphocytes.[6] It seems likely that lymphocytes in these organs cluster around a nerve terminal to receive a message and then depart, carrying the message with them. In a certain sense, lymphocytes behave like mobile neurons.

The cerebral neocortex, which can affect all the activities of the CNS, also seems to be involved in elaborating immune responses. In mice, lesions of the left cortex decrease T cell numbers and responses to an extent comparable to that observed after thymectomy, whereas B cells and macrophages are not affected.[7] In contrast, a similar right neocortical lesion does not affect the immune system. Thus, the neocortex appears to be involved in modulating the activity of the most elaborate arm of the immune system, the T-cell lineage. Moreover, this phenomenon is lateralized. Interestingly, the activity of imuthiol, an immunopotentiator that selectively enhances T-cell-mediated events, depends on an intact cerebral neocortex.[8]

Finally, it has been shown that an immune response can be behaviorally conditioned.[9] This phenomenon is well known in humans: asthmatic patients may have an asthma attack if they think the appropriate antigen is present (flowers, for example).

In conclusion, the hypothalamo-pituitary axis, the sympathetic nervous system, and the cerebral neocortex may all be implicated in elaborating immune responses. Questions remain to be answered as to whether this involvement is permanent or only occurs under extreme conditions and whether other CNS centers are involved.

Can the immune system "talk" to the CNS?

A basic requirement for postulating that the CNS is involved in immunoregulation is that there be a reciprocal flow of information between the immune system and the brain. The brain should be able to detect signals from the periphery and respond accordingly. Knowledge of how the central nervous system and the immune system could exchange information is still scanty. A number of results suggest that activated immunocytes may release messengers that affect neuronal activity.[10] The exact nature of the messengers is unknown, though immunological cells secrete compounds that can be recognized by neurons like histamine, serotonin, prostaglandins, and interferon.

Can immunocytes "understand" endocrine or neuronal messengers?

Immunocytes carry on their surface membranes receptors for hormones and for neurotransmitters.[11] Thus, immunocytes can "recognize" hormonal and neuronal messengers. Furthermore, some hormones and neurotransmitters can activate lymphocytes.[12] So it seems that immunocytes can not only "recognize" endocrine and neuronal messengers but can also modify their activity accordingly.

Conclusion

Today we know that certain brain lesions can alter an immune response, that the immune system can "talk" to the CNS, and that immunocytes can "understand" and respond to neuronal messages. Hence, the popular belief that our mental state may influence our immune responses may prove to be true. That the CNS and the immune system may both collaborate to maintain body integrity is hardly surprising when one considers that both systems have striking similarities. Both systems show a remarkable degree of cell diversity; both possess memory characteristics that do not exist in other systems; both employ cell-to-cell communication involving soluble messengers (lymphokines and neurotransmitters); and both systems are aimed at recognizing and reacting to the intrusion of foreign bodies to preserve species and individual characteristics.

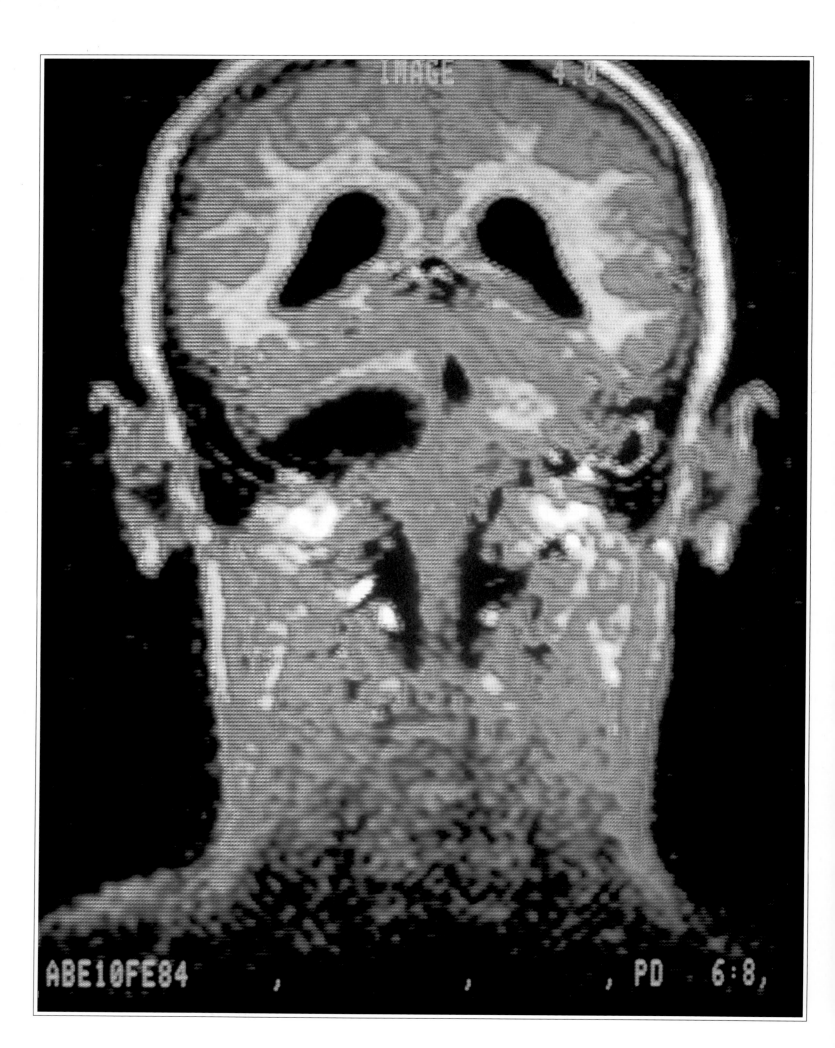

15
New Knowledge from Damaged Brains

For the mechanically minded, the temptation to take apart a broken machine is an overwhelming drive in childhood. Finding out what is wrong with a broken machine can be a very effective way to learn something. And the same thing is true with brains. Much of what we know about how the human brain works comes from studying what goes wrong when the brain does not work properly. Studies of the effects of stroke on language by Paul Broca and Carl Wernicke a century ago led to many insights, some realized only recently, about how the brain deals with language. This is an appropriate point in our story to consider how the study of damaged brains contributes so much to our knowledge of what the brain is doing. And like the child who learns to fix a television set by taking apart a broken one, we are learning a few things about fixing broken brains.

The classic case of Phineas Gage

Sometimes the lessons taught by broken brains are slow to be learned. A striking example in the history of brain research is the classic case of Phineas Gage. As a young man working on the railroad in New England in 1848, Gage was the victim of a freak accident. An explosion propelled a steel bar over a meter long and five centimeters in diameter into his left cheek and out through the top of his head on the right side. While this was not a particularly delicate piece of surgery, it did effectively disconnect the frontal lobes of Gage's brain. To everyone's surprise, Gage survived.

But was it really Gage, this person whose brain was no longer whole? Gage's friends seemed to think that Gage was no longer Gage. The Gage they knew had been a rather quiet fellow, polite and reliable, good at his job as a foreman on the railroad. The new Gage was none of these things. He was a loud, obnoxious fellow, impatient, obstinate, and capricious. His physician saw no loss of intellectual powers, although no tests existed at that time to probe them carefully.[1] Gage left his job and went on display in a circus for a time. The remainder of his life was occupied with purposeless wandering, and he ended his days in San Francisco in 1860.

This one instance of the effects of removing the frontal lobes is, of course, not conclusive evidence of the consequences of such an operation. But it should have served as a warning. Instead, the removal of the frontal lobes became a favored treatment for schizophrenia, as mentioned in chapter 7. During the 1940s and 1950s, tens of thousands of patients experienced often profound changes of behavior not unlike those exhibited by Phineas Gage. The procedure was justified on the grounds that it spared intellectual function while relieving the symptoms of the mental disorder.

But are the frontal lobes disposable? Their removal does indeed have little effect on the patient's ability to perform on standard tests of intelligence. But the frontal lobes have elaborate connections to almost all other parts of the brain. They develop rather late in life, in adolescence or early adulthood, when an individual's so-called higher mental capabilities begin to emerge: the ability to make elaborate plans and to pursue them with hope, the ability to relate in sensible ways to other individuals in different social situations. Without the frontal lobes there seems to be a dissociation between emotion and thought. Life is emotionally flat and actions are without thought.

The specific functions of the frontal lobes are not understood in the same way that we understand what the visual system does. And there are few tests to measure dysfunction of the frontal lobes. One test monitors where the subject is looking when a picture is placed before him. Normally, we look first at the face in the picture of a person, then scan other details in a fairly systematic fashion. The frontal-lobe patient has quite a disorganized pattern and directs his gaze almost at random. This has little practical effect on the patient's

Brain tumor
Computerized tomography makes it possible to see details of brain damage invisible in ordinary X-rays, like the tumor in the lower, left side of the brain.

life; it simply takes him a bit longer to interpret what he sees.

A more complex test is the Wisconsin modification of the Weigl card-sorting test. Each card has one to four copies of one of four different simple geometric figures in one of four different colors. The patient is asked to sort the cards into stacks, and the examiner tells him with each card whether it is correct or not. The criterion can be number, shape, or color. Frontal-lobe patients have little trouble doing this the first time, but when the examiner changes the criterion, they stick with their original scheme, no matter how many errors they make. They simply cannot change their program of action. We do not know what this test means. It is simply a pragmatic way to diagnose frontal-lobe damage.

In sum, the frontal lobes seem to be a kind of higher organization center. Their specific operations are still poorly defined, and we have only crude tests to probe their powers. Their functions, however, seem to be an important part of what makes us human. Today frontal lobotomies have largely been replaced by more delicate surgical procedures and powerful drugs that can be used to relieve psychotic symptoms in many patients without destroying the person in the process.

The two hemispheres: Facts and myths

One of the most widely publicized, and most widely distorted, stories of the study of damaged brains is the one that concerns split brains, brains whose left and right hemispheres have been separated by surgery or physical damage. This leaves two more-or-less-independent brain halves inside one person's head, and it is cause for a field day by imaginative writers, like those who claim that traditional schooling fails to educate the right hemisphere of our children. But what is the real story about what goes on in the two hemispheres? How do people with split brains really function, and what does their experience tell us about the day-to-day life of the two hemispheres in those of us in which they remain on speaking terms?

In looking at this work on the divided brain, we can see some justification for behaviorists discrediting introspection as a guide to how the brain works. The "conscious" self seems not to have access to much of the mental computation that takes place independently of our awareness. At the same time, we can see how radical behaviorists threw out the baby with the bathwater, because introspection proves to be an important tool for separating that still-elusive notion of conscious awareness from several very important aspects of what the brain is doing.

The roots of the story of split brains no doubt extend deeper into the history of neurology than anyone has yet been able to dig, just like the deepest roots of most scientific research. A good starting point is about 1940, the time of a sort of golden age of neurosurgery. A neurosurgeon in Rochester, New York, treated a number of patients suffering from epilepsy that defied other forms of treatment by severing the nerve fibers that connect the right and left hemispheres of the brain. This rather radical surgery was done with the intention of preventing electrical seizures that originated in one hemisphere from spreading to the other hemisphere. It seemed to work, sometimes. Most surprising, it seemed not to have any detectable effects on the patients' behavior.

Things got even more interesting when Roger Sperry, who later won a Nobel prize for his work, and his colleagues began to use the split-brain technique in animals to try to learn something about how the process of vision is shared between the two hemispheres. Because of the way our brains are constructed, everything to the left of the center of where we are looking is transmitted to the right hemisphere, while everything to the right of center goes to the left. The question on Sperry's mind was how do the two hemispheres share the information they receive to produce a unified perception of the world. To answer it, Sperry did some surgery not only to split the hemispheres, but also to cut some pathways running from the eyes into the brain to make sure that each brain hemisphere received information from only one eye. With this technique, he demonstrated a remarkable phenomenon. If a cat is taught that a particular pattern it sees with its right eye will reward the press of a lever with food, closing the right eye and opening the left eye produces an animal that has not learned. This experiment, which showed that learning in one brain hemisphere is apparently independent of learning in the other, provoked much of the nonsense about split brains that followed. Such experiments also paved the way, along a very difficult road, for much of the real understanding about the working of the mind that has emerged from the careful and patient explorations of the consequences of separating the hemispheres of the brain.

After carefully controlled animal experiments made it clear to observant scientists that the behavior of humans with split brains should differ in some detectable way from that of persons with intact brains, the time was ripe for some exciting new experiments. Many talented researchers arrived on the scene to try to find out what really goes on inside the head of a person with an interrupted link between right brain and left brain. One of the most careful was Michael Gazzaniga, who, as a young student of Sperry's, watched eagerly and thought hard as the early studies developed. It began with a patient, known by his initials W.J., whose brain hemispheres had been separated in a repeat of the uncertain efforts of 20 years earlier to treat epilepsy. After a quick flash of a picture that could register in only his left hemisphere, W.J. reported quite accurately and normally what he had seen. Yet he denied seeing anything that was flashed to his right hemisphere. Here were the first signs that something important might be learned by examining the effects of split-brain surgery in humans.

The story of split brains has now had years to mature. There is some sign that the surge of misinterpretation of the results has subsided. As people in marketing are well aware, people are always eager to buy an easy solution to difficult problems. For those working on split-brain research, the problem with traditional education lay in the failure to recognize the needs of the "repressed" and mute right hemisphere. As Gazzaniga has forcefully pointed out, though split-brain studies do indeed show that for most adults the two hemispheres tend to have different special capabilities, they definitely do not show that there reside within our heads two distinct minds locked in an endless cognitive conflict.

A more recent case of a split brain helps to set the stage for Gazzaniga's current view of what split-brain research tells us about the function of normal brains. This time the experiments involve a patient whose initials are J.W. By now the experimental techniques and the questions they address are much more sophisticated. J.W. faces a test screen and the word "bike" is flashed on it so that only his mute right hemisphere sees it. Asked what he

1-2. Epilepsy
Within the last five years epilepsy has become better understood. With the aid of living brain sections in laboratory dishes, it is now possible to provoke artificial seizures and observe their behavior. These sections of the hippocampus, the brain region most subject to seizures, are placed in a chamber with circulating fluid at body temperature. A microelectrode records the electrical activity of individual cells. Excessive electrical activity and other features characteristic of epilepsy are picked up and amplified 100 to 1,000 times for observation on a video display not shown here. Applying certain chemicals causes seizures, which can be used to test the effectiveness of antiepileptic drugs.

New Knowledge from Damaged Brains

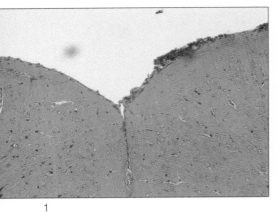

1

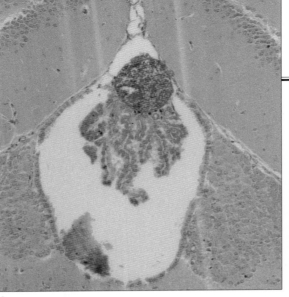

2

3

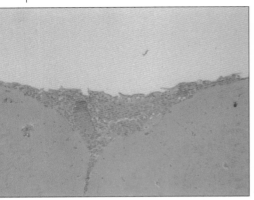

6

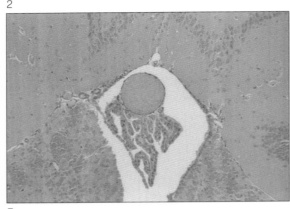

7

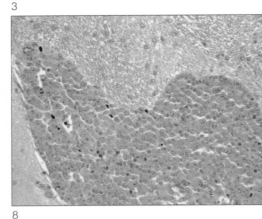

8

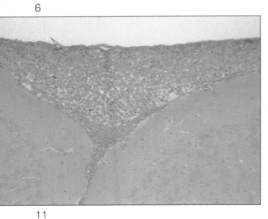

11

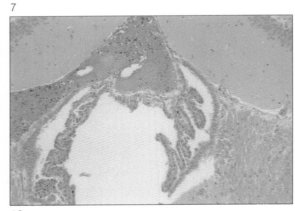

12

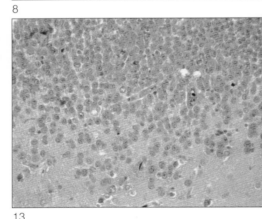

13

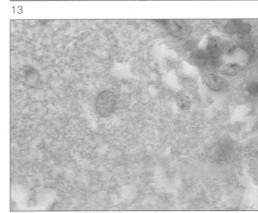

16

The brain and cancer
The cancers of the brain are generally malignant tumors of the protective membranes (ependymal epithelium) or of the supporting cells (glia), whose functions are poorly understood. True cancer of nerve cells in the brain is virtually unknown, because these cells do not renew themselves and are incapable of the rampant growth of cancer. The brain is protected by tightly joined cells that surround the blood vessels and prevent potentially toxic substances from entering into the brain (the blood-brain barrier). But the blood-brain barrier also prevents the entry of many anticancer drugs that could destroy the malignant cells. Surgery and radiation therapy are often the only resort. Finding a drug that destroys cancer cells without seriously harming healthy cells requires a laboratory-animal model of brain cancer to test candidate substances. One such model is illustrated here. With the aid of a microsyringe, the experimenter injects leukemic cells into an anesthetized mouse. On the day of inoculation, no abnormalities are observed in a cross section of the brain. The groove between hemispheres is normal (1). The ventricles have a normal volume, and the choroid plexus within them, which secretes cerebrospinal fluid, is normally irrigated by a large vessel (2). The meninges, which protect the surface of the brain, are normally organized (3), as are the arteries underlying them (4). The capillaries that supply the depths of the cortex are filled with red blood cells (5). Four days later there is an accumulation of abnormal cells in the interhemispheric groove penetrating the meninges (6). The choroid plexus shows signs of cellular alterations (7). Masses of leukemic cells proliferate under the inner membrane covering the brain (the pia mater) and push back the brain

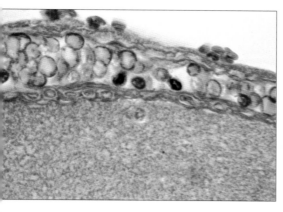
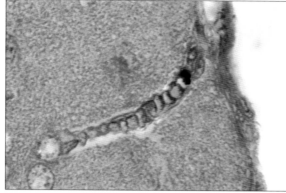

5

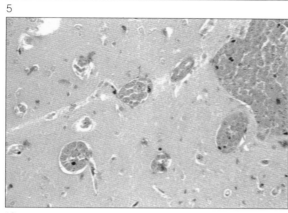

10

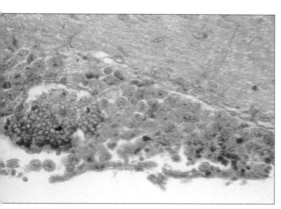
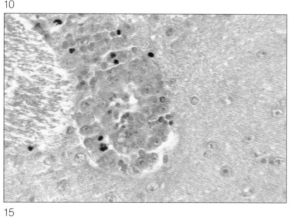

15

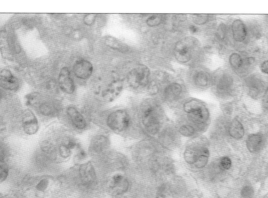

cells (8). Edema develops around peripheral blood vessels (9). Abnormal cells are also found around the penetrating capillaries (10). Eight days later microscopic examination of the brain tissue confirms the progress of the cancer's attack. A significant expansion of the cerebral volume has taken place during the evolution of the disease. The interhemispheric groove is now filled with cancerous cells (11). The mass of malignant cells compresses the brain. Serious changes occur in cerebral circulation. The irrigation of the choroid plexus appears modified (12). Under the pressure the leukemic cells burst through the most intimate protection of the brain and invade the brain tissue itself (13). The peripheral vessels disintegrate (14) or become obstructed (15) and open the way for a general dissemination of the disease. This pathologic state is shortly before the death of the animal. The glial cells become filled with holes and die (16), making room for more leukemic cells (17). During tests of new drugs that might penetrate the blood-brain barrier, the animal is treated from the day after the injection of the cancerous cells. If its survival is prolonged, the anticancer drug has directly reached the malignant cells in the brain.

New Knowledge from Damaged Brains

1. Malnutrition
Malnourished children like these clearly suffer physically from the effects of drastically limited diets. Nature generally assures that the brain gets preferential treatment, but we do not know with certainty how the brains of such children are affected by such grave malnutrition.

2. Coma
A nurse watches for the least sign of response from Janice during a session of sensory stimulation at the Greenery Clinic near Boston. Janice, a student of art and choreography in New York, has been in a coma for three years. She is one of the million comatose patients in the United States, a population that is increasing at the rate of a thousand per year.

3. Migraine
Long considered to be a largely imaginary affliction of women, migraine is today a serious subject of study for neurologists. It is often associated with disturbances of sensation involving flashing colored lights, zigzag lines, tingling sensations, or numbness. While there are many clues to its biological origins, like a dilation of blood vessels in the membranes surrounding the brain, migraine remains poorly understood to this day. (Engravings by George Cruikshank, 1835)

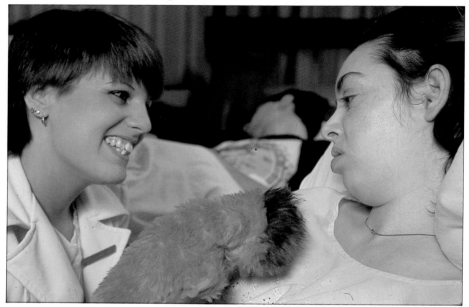
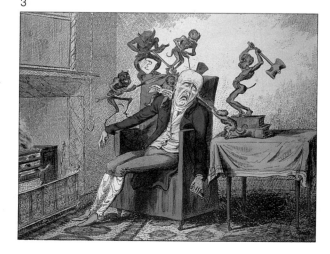

saw, J.W. reports that he saw nothing. But if J.W. is asked to draw what he saw with his left hand, which is under the control of the right hemisphere, the left hand will draw a picture of a bicycle. So there is some kind of language capacity in J.W.'s mute hemisphere, and a puzzling form of consciousness. Making something sensible out of such a discovery is a major scientific challenge. The interpretation that Gazzaniga and many other researchers apply to these and some additional puzzling aspects of human brain function involves a controversial concept known as the modularity of mind and brain.

What are the components of the brain?

The division of labor in the brain is a well-established principle. The auditory system analyzes sounds and the visual system analyzes the information that comes in through the eyes. Most researchers agree with this sort of modularity, the idea that the brain has different components to deal with different sorts of information-processing chores, just as a component stereo system has a record player, a cassette player, a tuner, an amplifier, and so on, each doing a different specific job. But recently scholars concerned with brain functions have pushed the idea of brain modules even further, an effort that, not surprisingly, has stimulated much controversy, as any new idea will do. At the moment modularity can account in principle for many sorts of phenomena that other views of brain function cannot explain very satisfyingly. As long as thinking about brain organization in terms of functional modules provokes useful new research questions, it is likely to remain alive and vigorous.

What are brain modules? The dictionary definition mentions independent functional units that form parts of a larger system. The operative word here, the one that lies at the heart of the controversy, is "independent." The implication that there are independent modules in the brain can be discomforting, since it implies a very definite stand on the question of consciousness. In short, it implies that the unconscious mind made so popular by Sigmund Freud may have a strikingly real physical manifestation in the brain; the module occupied by consciousness has the sometimes embarrassing responsibility of accounting for what one of its independent cohorts has come up with.

New Knowledge from Damaged Brains

The contrary view of how the brain works is, on the surface, a more comfortable one. Many distinguished scholars consider the brain to be a sort of general information-processing system that uses the same principles in listening to music as it does in looking at a painting, that constructs the grammar of language with the same tools it uses in building a bookshelf. Do different domains of the brain operate using their own laws, or is each wired to do a specific job? Research with split-brain patients, whose verbal memories are inaccessible to the nonlinguistic hemisphere, and ideas proposed by Ramachandran about the visual system's special bag of tricks argue in favor of modularity. As the frontiers of brain research advance, we will learn more about the validity of such arguments as those of Carnegie-Mellon University psychologist John Anderson, who favors a single coherent domain of mind, and of the philosopher Jerry Fodor, who favors modularity. We cannot predict what the future holds for these positions, but we can be fairly certain that neither is totally correct.

Karen Ann Quinlan
A story that captured the headlines throughout the world in 1975 was that of Karen Ann Quinlan, a 21-year-old woman who lapsed into a coma after an overdose of drugs. Doctors maintained her on a life support system for six months until her adoptive parents persuaded them to pull the plug.

(NY16-OCT.15)--COMATOSE--Karen Ann Quinlan, above, has "some hope of recovery" and should not be removed from a respirator according to her court-appointed guardian. The 21-year old woman's adoptive parents, Mr. and Mrs. Joseph Quinlan, are seeking to have their comatose daughter's physicians remove her from life sustaining equipment contending she is in a "chronic vegatative state" with no hope of recovery. She has been in a coma for six months. (AP WIREPHOTO)
(See AP AAA Wire Story)

(wt41024fls)1975
EDS...A 1975 file photo
(NY16-OCT.14)--KAREN QUINLAN--This 1975 file photograph of Karen Ann Quinlan is transmitted in response to member requests.
(AP WIREPHOTO)(wt31048fls)1975

New Knowledge from Damaged Brains

1

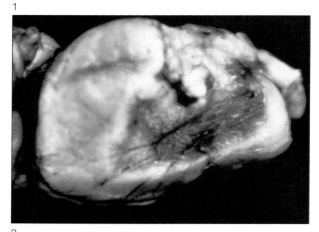

2

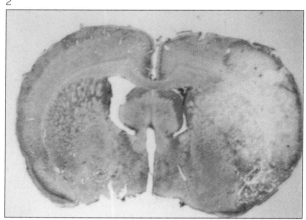

- Cortical edema
- Internal capsule
- Infarct of the striatum
- Rupture of the middle cerebral artery

Cerebral infarcts

A cerebral vascular accident (stroke) may appear as superficial bleeding but usually the interruption of blood flow results in the death of a region of brain tissue (the infarct). Tissue death begins within hours after the accident and continues during the following days. These figures show the results of occlusion of the middle cerebral artery in a two-year-old rat. The superficial bleeding is readily visible on the lateral surface of the brain (1). In a transverse section of the brain (2), the dead tissue normally perfused by the artery is readily apparent. The necrosis extends to the deep layers of the striatum, and the cortical region becomes swollen with accumulated fluid.

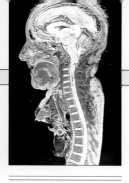

The Interpretive Brain

by Michael S. Gazzaniga, Dartmouth Medical School

It is hard not to think of personal consciousness as reflecting a unified, serially ordered process. Our thoughts seem to come into consciousness one at a time. They are considered, evaluated, and then put away so that we may entertain the next thought. When trying to solve problems, that is, to think in a concerted, rational way, we feel we move from one point to another in an orderly, serial manner directed by our language system. This concept about the nature of mental processes has been formalized for years as the main view of experimental psychology. There has also been tacit approval of this idea from the brain scientist. Hierarchical models of brain function are commonly proposed when simple percepts are managed by neurons early in the information processing sequence. Subsequently, the messages processed by these first neurons receiving sensory information from the environment become molded and abstracted into more and more complex messages that reflect more complex perceptions.

This overall view of a unified, serially ordered system is being challenged in modern cognitive science and neuroscience. In cognitive science what once looked to be unitary mental activities are not. They are made up of constituent structures that are active in overall mental activity but identifiably separate. Each can be examined and its characteristics described. Thus, human thought, once believed to be a unique product of our species' capacity for language, is now viewed as the product of a constellation of parallel processes located in both functionally and physically separate modules. These modules, most of which work outside the realm of consciousness, are continually active in processing information that is crucial for any mental activity. They work in parallel; that is, they are cocontinuous with the process by which information is admitted into our consciousness. The same view is also arising out of the brain sciences. For example, studies on animal brains reveal there is not just one brain area involved in vision but several. One area may be involved in motion detection, another in color detection, another in brightness detection, and so on. In humans, studies on patients who have undergone hemisphere disconnection to control otherwise intractable epilepsy, so-called split-brain patients, have emphasized this view and have also supplied insights into the critical human brain mechanism that subserves our belief that we posses unified consciousness. This capacity is the product of a special module we possess in the left hemisphere called the interpreter.

Split-brain patients revealed this property of human cognition because of the special consequences of their surgery. They have the neural connections between their two half brains interrupted to prevent the interhemispheric transfer of seizure activity. This also prevents the transfer of sensory and motor information, thereby leaving each half brain functioning relatively independently from the other. Studies have shown information presented to the right hemisphere cannot be verbally described, even though it can be responded to in a nonverbal way. Thus, if a picture of a spoon is presented to the right brain, the subject says he sees nothing. That is the disconnected, left brain talking, since only it is capable of speech. However, studies have shown that the right brain in some patients is capable of directing the left hand in a search through a grab bag full of objects and placed out of view.

Typically, the left hand correctly retrieves the appropriate object, the spoon. Nonetheless, the patient, holding the spoon in the left hand, still is unable to say what it is, because the speaking module is in the left brain, and the right brain solved the problem.

Hundreds of experiments have been carried out on these patients, each emphasizing the specific ways the brain is specialized for cognitive activity. But perhaps the most telling is the discovery of how the left-brain interpreter works to synthesize the behavior and thought that can be produced by the separate modules. The classic demonstration is to give a simple problem to each hemisphere to solve. Because of how the visual system is organized in the brain, information presented to the left of a person's field of view goes to the right hemisphere, and information presented to the right goes to the left hemisphere. In a test on patient P.S. the right hemisphere viewed a snow scene and appropriately picked a snow shovel as the appropriate match, while the left viewed a chicken claw and picked the chicken. After such a response the split-brain subject, who can only talk out of the left brain, was asked by the examiner, "Why did you do that?" P.S. immediately said, "Oh, that's easy: the chicken claw goes with the chicken and [looking down at his left hand pointing to the shovel] you need a shovel to clean out the chicken shed." In short, the left-brain interpreter, observing behaviors produced by some of its separate modules, constructs a theory that fits why such behavior might have occurred. It does this instantly, smoothly, and matter-of-factly.

This illustrative example summarizes much of our new knowledge about how the human brain is organized to generate human thought and beliefs. Brain states managed by separate modules allow independent production of both overt and covert behaviors. In the human these activities are not left as capricious events. A special interpreter views all actions and thought emerging from the vast array of modules and generates hypotheses about why we do what we do. These, in turn, become our beliefs, our own special, personal view of life.

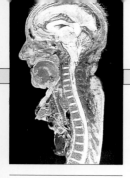

Electrical Signs of Lateral Dominance in the Human Brain

by Richard Jung, Albert Ludwig University of Freiburg

For more than a century neurologists have known that the left human forebrain is specialized for language. This language dominance of the left cerebral cortex was deduced by Broca and Wernicke in 1865 to 1873 from their observations of speech disturbances following unilateral brain damage. An opposite right-sided spatial dominance was postulated by Holmes and Reichardt in 1916 to 1918 on the basis of their observation that visuospatial defects were associated with damage to the right cortex. Until recently no objective signs of these cerebral lateralizations for language and space perception were observed in normal people. In 1982, however, my colleagues and I began recording slow event-related brain potentials from normal people and looking for such objective signs. Since then we have found left-sided electrophysiological correlates of dominance for language and calculation and of right-sided dominance for spatial cognition.[1] Here I shall summarize our main results and discuss cerebral dominance in right- and left-handers.

Language, calculation, and spatial orientation are learned functions acquired in childhood after many years of practice. Hence, recording their cerebral correlates may enable us to investigate the development of these higher brain functions.

Methods

We recorded and averaged slow brain potentials from 10 cortical regions on the scalp of normal subjects while they performed linguistic and calculating tasks, visuospatial tasks, or learned movements. During the tasks, which last 10 to 15 seconds, eye movements and vocalizations were forbidden, to avoid electrical artifacts. A normal population of 120 right-handers and 60 left-handers was studied. Three periods were analyzed: a control period of eye fixation, a task period of solving problems or viewing figures in space, and a writing period for writing down task results.

Language was tested by word completion, association, or finding synonyms; calculation by the addition of two-digit-numbers. The visuospatial tasks consisted of viewing ambiguous figures that allow reversal of perspective (such as the Necker cube), recognizing stereoscopic patterns, or mentally transposing two-dimensional figures into three-dimensional ones.

Recordings made from the left and right hemispheres during the performance of various tasks display bilateral electronegative potentials. The surface-negative amplitude of these potentials is larger than that of readiness or expectation potentials, which may appear during the preceding control period of fixation. When solving the language problems, most right- and left-handers show larger negativity over the left hemisphere. These left-sided lateralizations of bilateral negativities are most pronounced over the frontal brain.

When normal right- and left-handers perform both language and calculation tasks, left-sided lateralization occurs in the majority of both groups, but it is more evident in linguistic than in calculating tasks. The rare right-sided lateralizations occur more often in left-handers than in right-handers. Among 100 right-handers tested for linguistic tasks 77 showed left-sided lateralizations. Only 8 right-handers showed right-sided lateralization for the same task.

Among the 50 left-handers tested in linguistic tasks, 33 showed a predominantly left-sided lateralization, and 11 a predominantly right-sided lat-

Left-sided dominance of language and calculation in right- and left-handers

205

eralization during language tasks. This is inconsistent with an often-quoted postulate of Broca in 1865: "Left-handers have right-handed brains," but it agrees with neurosurgical and neurological evidence for left-sided speech dominance in left-handers.

During calculation tasks, a left-sided lateralization also appeared in most right- and left-handers. However, it was less frequent and smaller than during language tasks. Bilateral negativity with little lateral bias occurred more often (in 37 percent of the right-handers and 22 percent of the left-handers) than in the language tasks.

Right-sided dominance of spatial perception

Clear right-sided lateralization was recorded for spatial tasks in the same right- and left-handers who showed left-sided lateralization for language and calculation. We had expected from neurological experience that complex spatial tasks would produce more lateralization to the right parietal cortex than would the simple tasks associated with ambiguous perspective or binocular stereopsis. To our surprise, however, right-sided lateralization turned out to be most frequent and well localized over parietal and temporal regions during the simple tasks. This indicates that a bilateral coordination of stereoscopic depth signals from both eyes in the occipital areas, as shown by bilaterally equal visually evoked potentials recorded by Julesz and coworkers, is followed by *unilateral* spatial processing in the right parietotemporal cortex. The right-sided lateralization in perspectival and stereoscopic perception was independent of handedness. In complex spatial processing, however, females showed right-sided lateralization more often than males. We assume that the females used more visual cues for problem solving than males, since some male subjects used logical procedures to determine the two-to-three-dimensional transformation.

Viewing the spatial reversals of ambiguous Necker figures results in right-sided lateralizations during the task. When subjects view stereoscopic random dot patterns, perception of depth usually occurs during and after the first second of stimulus exposure. Since electrophysiological lateralization also develops during the first second of stimulus projection, this supports our hypothesis that the slow potential lateralization correlates with higher processing of spatial information in the right brain.

During linguistic tasks and calculation, left-sided lateralization is rather widespread over the anterior forebrain from the frontal to the parietal and temporal regions, including the precentral motor cortex. This contrasts with the spatial processing task, which shows more localized negativity over the posterior cortex of the right hemisphere.

In conclusion, cognitive processing of language and space perception are accompanied by slow brain potentials showing unequal but bilateral cortical activation. Hence, the terms "lateralization" and "dominance" can be used only in a relative sense. The electrical signs of relative cerebral dominance appear for language and calculation in the left hemisphere and spatial perception in the right hemisphere of the normal forebrain. These lateral biases of surface negative potentials are more widespread than predicted by neurological observations of brain lesions. They are always superposed on activity in both hemispheres. The extension of the potentials over both forebrains demonstrates that many cortical areas participate in these higher functions of language and cognition. The module principle, proposed by Szentagothai and others,[2] by which the functions of many cortical columns can be coordinated, may eventually explain how special interactions of the two hemispheres and their callosal transfer can develop with learning into unilateral specializations and dominant functions of the cortex.

Our electrical recordings are paralleled by other measurements of brain metabolism and blood flow in the cerebral cortex during speech and tactile space exploration. Isotope imaging, used by Ingvar and others, has also demonstrated bilateral activation during speech, with some unilateral bias in the left frontal and temporal cortex.[3]

Unilateral dominance, specialization, and learning

Obviously language, calculation, and space orientation are special functions of the brain, acquired by learning, and fixed in memory. They develop during many years of practice in childhood. Although doubtless based upon some innate structural order, the relative dominance of unilateral brain localization is dependent on experience.

Learning and brain potentials

Since the developmental correlates of cortical functions are unknown, it may be useful to study simpler cases, like the learning of movements, and related slow potentials. The short-term learning of motor action during repeated practice is accompanied by a diminution of widespread slow potentials in the cortex but also by a rise of potentials in the specially activated cortical areas.[4] Investigations of these phenomena are only beginning, and further studies are clearly needed.

Although the mental functions of language and cognition have been acquired during many years of learning, the brain correlates of learning are virtually unknown. How the slow potentials develop during children's learning of language and writing remains to be investigated.

CICERON

16
Speaking Your Mind: What Is Language?

Everyone knows what language is, of course. We use it all the time. A language consists of a set of words, the vocabulary, and rules for combining them, the grammar. For centuries everyone agreed that language is solely a human attribute. Humans are distinguished from animals by their use of language. Then in recent years, people began teaching chimpanzees and gorillas to use sign language and the distinction began to blur. After all, there is only a one or two percent difference between human DNA and chimpanzee DNA. How much difference can this small percentage make between our brains and chimpanzee brains?

Linguists and psychologists are still arguing over whether apes taught to use sign language really do have mastery over language in the strict sense. Some argue that since the apes do not have full command over the rules of grammar, they have not really learned to use language. Others assert that since the animals can create new terms to describe previously unencountered situations, they display an essential feature of human language use. One chimpanzee, for example, produced the signs for "water" and "bird" together on seeing a duck in water for the first time. But is this a true invention? Most of these arguments seem to generate more heat than light and are not interesting to explore here.

The origins of language: Wild children

Putting aside the rather frustrating debate about whether apes can use language, it is clear that they did not independently invent anything like a human language. Where then did language begin, and how? This is a question that people have asked since antiquity. The Greek historian Herodotus reported a supposed experimental attempt to answer it that he heard about during his travels in Egypt about 460 B.C. As the tale goes, the Egyptian king Psammeticus wanted to see which language would naturally emerge in children raised without any contact whatsoever with an existing language. (His actual concern was to determine whether Egyptian or Phrygian was the "original" language.) He sent two children to be raised by a shepherd who was instructed to treat them well but be sure that they never heard a spoken word. At the end of two years of this treatment, the children one day came running to the shepherd crying "becos." Since "becos" is the Phrygian word for bread, the good king had to concede that Phrygian preceded Egyptian in its origins.

This charming little fable tells us little about what the first human language might have been like, and ethical considerations make it impossible to repeat the experiment of king Psammeticus today. There have, however, been several modern instances of children raised without exposure to language, not as an experiment, but as the consequence of misfortune or mistreatment or both. The most famous of these are so-called wild children, children raised by bears or wolves. Harvard University psychologist Roger Brown made an extensive study of such children some years ago and provided a record of his findings.

One of the earliest cases on record was the discovery in India in 1920 of a wolf mother with two cubs and two human children, one about eight years old, the other some 18 months old. No one knows how the humans came to be there, but they were decidedly wolflike in appearance and behavior, favoring a diet of raw meat and eating by lowering their heads to the food. The younger boy died within a year of his discovery, while the older one survived about ten years. But the latter never fully assumed human capabilities; he could speak only a few words. The same has been true for the several dozen other instances of wild children. None was able to master language use.

What do wild children tell us about language? Some interpreters proposed that such children demonstrate only those human capabilities that are innate. Since they lacked

Cicero
The foremost orator of ancient Rome, famed for his mastery of language.

Speaking Your Mind: What Is Language?

1. Reading machines
Computer technology makes it possible today to build machines that can read books to people who cannot see.

2. Theater for deaf-mutes
The study of sign language as a means of communication and self-expression by congenitally deaf persons has helped us to understand how the human brain is organized to produce and understand language.

language, this view holds, language is solely a product of the environment. Others argue that the children were retarded in the first place and perhaps were abandoned for that reason. Or they may have been adversely affected by their unusual diet or the psychological stress of their unusual environment. But as Brown points out in his analysis, you can have it either way. These were not carefully controlled experiments. Too many factors are unknown. Psychologists Lila and Henry Gleitman at the University of Pennsylvania conclude from their analysis of wild children, "All we can say for sure is that having a wolf or a bear for a mother is not conducive to learning human language."

It is possible to learn a bit more about the biology of language by studying, not the extreme cases of wild children, but rather those of children who have, for various reasons, been kept out of contact with human language. One famous case is that of a child known as Isabelle, whose mother was a deaf-mute. Isabelle was kept out of contact with the outside world until she was discovered at the age of six. At that time she was unable to speak and had the intelligence of a child less than two years old. But after entering a school class with children her own age who had been using language for some years, it took her only a year to catch up.

A linguist at the University of California, Los Angeles, Victoria Fromkin, reported a more recent case of a girl named Genie, found in California at the age of fourteen. Despite intensive professional training, Genie never gained complete mastery over language. She uses largely childlike expressions even to express ideas far more complex than a child could. She lacks the ability to use elements that provide the fine texture of language, like auxiliary verbs and pronouns.

These two examples, together with other evidence, suggest that language must be learned before a certain age if it is to be mastered completely. Once a certain critical period has passed, the ability to learn language declines rapidly. Supporting this conclusion is the observation of how readily young children learn a second language, a talent that is the envy of many an adult preparing to travel abroad. There is also clear evidence that young children may recover remarkably well from brain damage that disrupts language use. Adults cannot recover nearly as completely, if at all. The accumulated evidence all suggests (but does not prove) that the biological stage in the human brain is set for language learning only during the years prior to puberty. After that period, the curtains are drawn.

The structure of language

While Noam Chomsky's name is not a common household word, it is surprisingly widely known for a name attached to rather arcane research on language. But Chomsky did give us revolutionary ideas in a monograph published in 1957 with the innocuous title *Syntactic Structures*. In its strongest form, Chomsky's notion is that all languages share a universal grammar, one that lies below the surface of all the different grammars for French, English, Japanese, and other languages. This line of reasoning also leads to the idea that the human brain contains a distinct language organ specialized for processing language, just as the heart is specialized for pumping blood and the stomach for digesting food. This proposal has profound biological implications for how the brain works, and it fired a still unsettled controversy: does the brain contain specialized cognitive organs for language, vision, mathematics, and so on, or does one general-purpose computer perform all the different jobs?

The unspoken language

Chomsky's work tends to be quite abstract and has frustrated many attempts to apply it to real-world use. Nonetheless, it has influenced the thinking of many researchers. One fruitful line of research that has a bearing on Chomsky's ideas focuses on the unusual form of language used by members of the deaf community.

Most people take for granted that language means speech. That is not necessarily so. This lesson has been learned slowly and laboriously in recent years by research on the use of sign language by deaf persons. One group that has been at the forefront of this research is led by Ursula Bellugi and Ed Klima at the Salk Institute in San Deiego, California. They have demonstrated that American Sign Language is a visual language constructed from movements in space. It has an elaborate grammar of its own and is every bit as complex, subtle, and flexible as spoken language. And because it is delivered by movements of the arms and hands in space rather than movements of the vocal apparatus and is perceived by the visual system of the brain rather than the auditory system, it offers wholly new insights into language-processing systems in the brain.

One of the most surprising discoveries that this work reveals has to do with the specialization of the brain's hemispheres for language and spatial analysis. In most individuals the left hemisphere governs language use, while the right hemisphere governs spatial perception. These specializations are vividly demonstrated in brain-damaged patients. Left-hemisphere damage can impair the ability to use spoken language, but leave unaffected the ability to perform spatial tasks like copying a picture of a house or putting together colored blocks to match a pattern. Right-hemisphere damage reverses the situation.

What effects might brain damage have on the language capabilities of deaf persons who have used sign language all their lives? Bellugi, Klima, and their associates located a number of adult sign-language users who had been deaf from birth and had recently experienced some form of brain damage. The scientists tested both language capabilities and performance on standard tests of spatial perception. Those individuals with right-hemisphere damage had difficulty on the visual tests as expected, but their use of sign language was virtually unaffected despite these problems. On the other hand, individuals with left-hemisphere damage did well on the visual tests but had varying degrees of difficulty using sign language. These results lend support to the concept of a specialized organ for language, whether spoken or signed, located in the left hemisphere of the brain, as some language researchers have taken Chomsky's work to imply.

1-2. Marilyn Monroe sings
While under the skin they are obviously machines, robots can be dressed up and made to look and act quite lifelike. This example, constructed at Waseda University, masquerades as Marilyn Monroe singing "River of No Return."

3. The tower of Babel
The Lord said, "Here they are, one people with a single language, and now they have started to do this; henceforward nothing they have a mind to do will be beyond their reach. Come, let us go down there and confuse their speech, so that they will not understand what they say to one another" (Genesis 11).

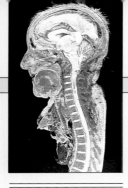

Developmental Dyslexia

by Albert M. Galaburda, M.D., Harvard Medical School and Beth Israel Hospital

"Developmental dyslexia" refers to a group of conditions all of which have in common the fact that the affected individual experiences unexplained difficulties in learning to read and write.[1] In other words, there are no problems with intelligence, emotional balance, or socioeducational opportunities. The popular concept about what dyslexia represents is likely to change, because it can now be considered in the light of new knowledge on the anatomy of the brain in dyslexia. Developmental dyslexia can no longer be viewed as an isolated educational or psychological issue. The purpose of this brief review is to bring readers not familiar with these late developments up to date with the current thinking on the problem of developmental dyslexia.

Early in the history of the study of dyslexia it was believed that the learning deficit was probably secondary to developmental abnormalities in the brain. Thus, it was thought that areas in the adult human brain believed to subserve reading and writing were improperly developed in the dyslexic individual. However, as anatomical evidence for this claim was not forthcoming, neurological theories were gradually abandoned, and in the 1920s and 1930s they were replaced by psychoeducational explanations. Research up until the most recent past concentrated on the development of diagnostic tests and general psychological characterizations, both of which contributed significantly to the establishment of educational therapies aimed at correcting the written-language deficits.

In 1979 Galaburda and Kemper reported brain findings after postmortem examination in a young man with developmental dyslexia.[2] These authors found evidence of disturbance in the prenatal development of the cerebral cortex, especially in the parts of the left hemisphere that are associated with language processes. Subsequently, Galaburda and colleagues studied the brains of several other individuals with developmental dyslexia, and the results of these studies confirmed the original findings and led to the development of neurological hypotheses that link the presence of the language disorder to developmental brain abnormalities and to other medical conditions.[3]

The development of the mammalian brain undergoes several steps that include the generation of neurons and glia, the migration of these cells to their final resting places, and the maturation of neurons, their interconnections, and their chemical characteristics.[4] During the last stages of brain maturation, a period that begins during the second half of gestation and continues probably into the first decade of life in the human, there is a systematic elimination of neurons, connections, and some of the chemical repertoires, elements that were originally overproduced and are then selected out and eliminated mainly on the basis of the genetic and environmental characteristics and requirements of the individual. In the human brain, neuronal migration to the cerebral cortex occurs approximately between the sixteenth and the twentieth week of gestation, and maturation begins soon after neurons have arrived in the cortex.

Injury to the cortex (by physical trauma, virus infections, X-ray irradiation, circulatory disturbances, etc.) during the period of neuronal migration in experimental animals has been shown to lead to ectopias (nests of cells that are malpositioned in the mature cortex) as well as to dysplasia (alteration of the number of neurons and the architecture of neuronal circuits surviving the systematic elimination) (figure 1).[5] In every one of five consecutively studied dyslexic brains, we found ectopias and dysplasia in the cortex surrounding predominantly the left Sylvian fissure (a region of the brain that contains the classic language areas) (figure 2). In routine autopsy examinations these findings are rare. We concluded, therefore, that events causing injury roughly at midgestation could be responsible for the ectopias and dysplasia seen in the dyslexic brains. Current research activities are being aimed at disclosing the exact cause or causes of the injury and its detailed anatomical and behavioral consequences.

Several studies have demonstrated that when injury to the developing brain occurs at specific developmental stages, it leads to significant reorganization of the cortical architecture and its pattern of connections not just locally in the area of damage but also in distant, often connectionally related areas.[6] The best interpretation of these changes is that the injury itself or chemical substances released by the injury interfere with neurons and axons surviving during the final stages of corticogenesis. Whereas some parts of the cortex clearly show fewer surviving neural elements, other parts show clearly exuberant development in terms of both excessive numbers of neurons and aberrant connections. The relevance of these processes to dyslexia will be discussed below.

Over 65 percent of human brains exhibit leftward asymmetries in language areas.[7] In other words, a given language zone is more often than not found to be larger on the left side than on the right. This is true, for instance, for the planum temporale, an area comprising a portion of Wernicke's speech area, located on the upper surface of the temporal lobe. The odds of finding a brain with a planum temporale that is symmetrical in size is on the order of one in eight cases. Nevertheless, every one of seven consecutively studied dyslexic brains had a planum temporale that was symmetrical, an extremely unlikely situation (figure 3). It is possible, therefore, that the presence of symmetry in a language area represents a structural accompaniment of developmental dyslexia. The co-occurrence of two unusual events suggests that the cortical anomalies (ectopias and dysplasias) and the symmetry are linked to one another. Thus, perhaps early injury causes reorganization of cortical architecture in a way that gives rise to symmetry rather than asymmetry. Symmetry in this particular case would imply improper wiring of the language substrates and consequently abnormal language-processing ability.

In sum, in developmental dyslexia there seems to be critical, early, focal injury to the developing cortex that leads to the inappropriate survival of neurons and connections, with resultant changes in the patterns of organization of the language areas. The anomalous cellular architecture and aberrant connections can then be summoned to explain abnormal language function.

Research is currently in progress to ascertain whether subtle prenatal injury (such as that producing ectopias of cortex) is capable of causing a reorganization of the type described above. While this is going on, another research effort is directed at trying to find out possible causes for the injury in dyslexics. Some clues may be available from the recent work of the late Norman Geschwind and colleagues.[8] This research demonstrates more learning disabilities and disorders affecting immunological function among individuals with anomalous cerebral dominance (left-handedness and ambidexterity). Furthermore, several strains of immune-defective laboratory mice exhibit cerebrocortical ectopias and dysplasias and abnormal learning behaviors.[9] Perhaps prenatal brain injury caused by immunological lesions produces ectopias and dysplasia and serves to trigger the reorganizing effects that lead to anomalously formed language areas.

While many of the characteristics of normal and abnormal brain development mentioned in the preceding paragraphs have not been specifically tested in learning-disabled individuals, enough evidence is already available to remove dyslexia from the category of a psychological or educational problem and to open new avenues of research. The next few years, I hope, will see a clarification of the biological characteristics of dyslexia and the emergence of treatment strategies for severely affected individuals who do not respond to currently available educational therapies.

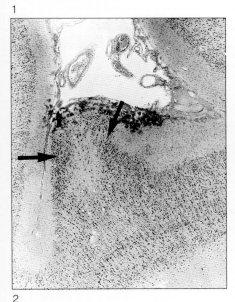

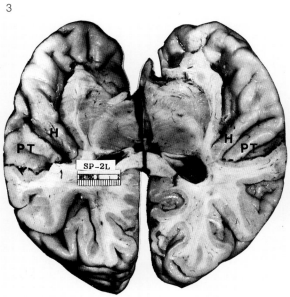

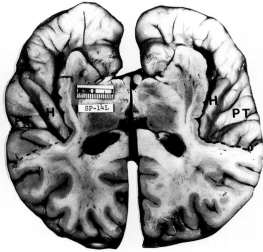

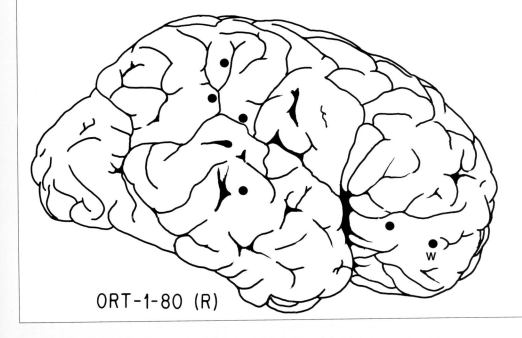

1. Ectopia and dysplasia in the brain of a dyslexic
Photomicrograph of a section of human cerebral cortex from the brain of a dyslexic individual to show ectopia (outlined by arrows) and distortion of cells (dysplasia) in the center of the ectopia.

2. Distribution of ectopic-cell growth and displasia
A diagram of the left hemisphere (above) and right hemisphere (below) of the brain of the dyslexic individual shown in figure 1 to show the placement of ectopias and dysplasias of cerebral cortex (dots). Note that the perisylvian cortex of the left hemisphere is more severely affected than that of the right.

3. The planum temporale
Photographs of the upper surfaces of the temporal lobes in two brains to show the planum temporale (PT). The brain on the left shows the typical pattern of asymmetry in the size of this structure, whereas the one on the right (from the dyslexic individual of figures 1 and 2) is symmetrical. (H represents the auditory gyrus of Heschl.)

215

A Drug for Dyslexia

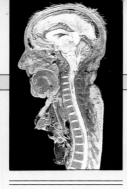

by C. K. Conners, Childrens Hospital, National Medical Center, Washington, D.C.

The electroencephalogram (EEG) is a recording of the brain's electrical activity picked up by metal electrodes applied to the scalp. The minute signals of ten millionths to a hundred millionths of a volt that reflect the activity of large groups of brain cells are amplified some millions of times to deflect movable pens that trace the EEG on a moving strip of paper. The changing details of these traces reflect changes in brain activity like those that occur during different levels of alertness and sleep.

To help interpret the information contained in the EEG, neuroscientists often analyze the amount of power in the signal that falls within different frequency bands. By tradition, the human EEG is divided into four frequency bands: delta at 0.5 to 4 hertz (cycles per second), theta at 4 to 8 hertz, alpha at 8 to 13 hertz, and beta at 13 to 25 hertz. Low voltage beta activity is associated with alertness, while slow delta waves appear when a person first goes to sleep. These characteristics of the EEG make it useful for analyzing the effects of drugs on brain function. The experiments illustrated here demonstrate the effects on brain activity of the brain-stimulating drug piracetam, which improves the reading capabilities of some dyslexic children.

To make the colored maps of brain activity, 22 electrodes distributed over the scalp deliver signals through a set of filters that decompose the signals into different frequency bands. The power in the different bands is then mapped onto a profile of the brain, which provides a general sense of which brain regions are producing relatively more or less power in a given frequency band. The color coding uses red for the highest power level and yellow, green, and blue for decreasing power levels. Our experiments show that piracetam reduces the slow delta waves, while having little effect on fast beta waves. This result suggests that the drug reduces brain activity associated with sleeping.

Event-related potentials (ERPs), another type of brain electrical activity, provide further information about the effects of piracetam. ERPs can be recorded from the scalp within a fraction of a second after the presentation of a visual or auditory stimulus. Certain components of the ERP reflect changes in attention or decision-making processes. In the experiment shown here, a dyslexic child saw a new letter of the alphabet every 1.5 seconds and was required to press a button when an occasional letter, termed a target, was the same as the one before it. The amplitude of a component of the ERP, the P300, which occurs about 300 milliseconds after the presentation of a stimulus, is considered to reflect the amount of information conveyed by the stimulus and is larger when a target occurs.

In experiments by my coworkers and me the administration of piracetam increased the amplitude of the P300 to targets and decreased its amplitude to nontargets. (Maps of P300 amplitude are made in the same way as the maps of power, with red representing the highest amplitude in this case.) These results suggest that piracetam makes the brain better able to discriminate between meaningful and meaningless stimuli.

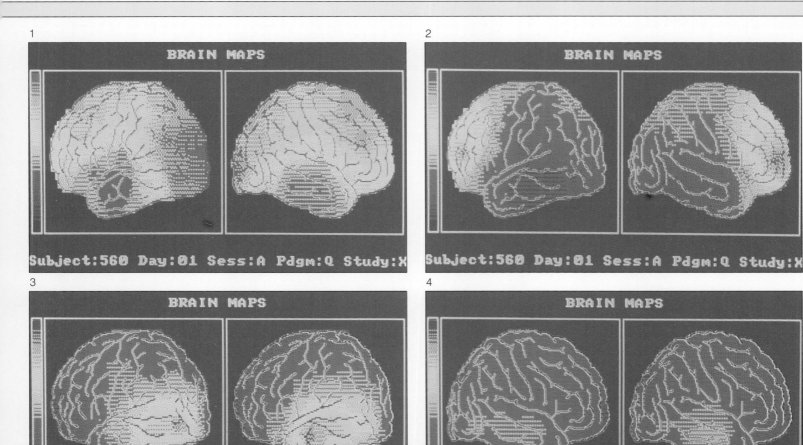

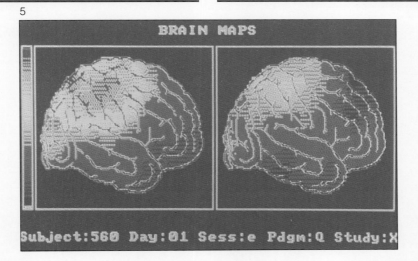

1-2. Effects of piracetam on delta waves
View of left- and right-hemisphere delta-frequency-band activity recorded from the scalp of a dyslexic boy before (1) and after (2) treatment with piracetam. Before taking the drug, delta-wave activity is high in the occipital region, particularly in the left hemisphere, suggesting that the brain is in a resting state, especially in the left parietal and occipital areas. After the drug was taken, the delta activity is largely restricted to the frontal regions. In these and the following two figures the levels of activity, from the highest to the lowest, appear as red, yellow, green, and blue.

3-4. Effects of piracetam on beta waves
View of left-hemisphere (3) and right-hemisphere (4) beta-frequency-band activity recorded from the same boy before (left panels) and after (right panels) treatment with piracetam. This activity is little affected by the drug. The contrast in beta activity between the two hemispheres is striking.

5. Effects of piracetam on event-related potentials
In this experiment, the dyslexic boy was required to press a button when one of a sequence of letters of the alphabet was the same as the one before. The brain maps show the distribution of the amplitude of the P300 component of the event-related potential, which occurs 300 milliseconds after presentation of a letter that did not require a button press, before (left) and after (right) treatment with piracetam. The lower amplitude after treatment suggests an improved ability to ignore meaningless stimuli.

17
Focusing In: Attention and the Brain

"Pay attention!" is a command that we heard all too often as children. We all know what it means, even though it is something that we do inside our heads. There may be outward signs of paying attention, particularly in the eyes. The eyes may shift position to focus on something we look at or lose the glazed-over look of a daydreamer. We can also accomplish such feats with attention as picking out one conversation at a noisy cocktail party. Recent research has also confirmed that we can shift our attention around in our field of vision to catch what is going on in a corner of our eyes while apparently staring intently at something else. We experience attention as a filter that our brains apply to the flood of information pouring in from the senses, a filter that removes everything but what interests us at the moment. Indeed, the modern research on attention began in the 1950s with the formulation of the filter theory of attention by British psychologist Donald Broadbent. According to this theory, the brain has only a single channel, like a telephone line, that connects sensory information to consciousness; paying attention is like connecting this telephone line to a particular sensory input channel.

In the years since Broadbent first formulated it, filter theory has been revised and refined a number of times. One of the major changes has been the discovery that a certain amount of processing goes on before we are even aware that anything is happening. One of the best ways to demonstrate this preattentive activity is in an experiment conducted by Anne Treisman of the University of British Columbia. If you show someone a picture containing a dozen or so small green circles jumbled together with only one blue circle of the same size, the viewer will effortlessly pick out the blue circle in a fraction of a second. Furthermore, the time it takes to perform this preattentive task will not increase significantly if you add more and more distracting green circles, which shows that the viewer is not searching through the circles one by one to identify the target. This pop-out phenomenon also works with other differences, like finding a few horizontal lines in an array of vertical lines. There is evidently a process that directs our attention toward a part of a scene that is interesting in some way.

Seemingly minor changes in the nature of this task dramatically change performance. If, for example, the assignment is to find a green T in a mixed array of green Xs and red Ts, it takes considerably longer than the tasks described above. This time the length of time it takes to find a target increases directly with the number of distracting items added to the array. This result suggests that in looking for a conjunction between two attributes, such as color and shape, we must focus our attention on each item in the array one by one to determine whether it has all of the required attributes. Treisman claims these findings as evidence that the brain's specialized systems for responding selectively to such different attributes of visual stimuli as orientation, color, stereoscopic depth, and movement are capable of preattentive awareness of each attribute but that awareness of a conjunction of different attributes requires focusing what she refers to as the "searchlight of attention" on the spatial location of each object.

While the behavioral aspects of attention readily offer themselves for study, the functions of the brain associated with paying attention are much more difficult to identify. Behavioral studies have revealed that if we are told to prepare to hear a tone of a particular frequency or see a light flash at a particular location in space, we become significantly more sensitive to whatever it is we have devoted our attention to. That is, we respond more rapidly and with a lower threshold to stimuli with the feature for which we are prepared than we would if we were not paying attention. It seems likely, therefore, that the sensory channel in the brain that is activated by attention should show signs of being more responsive. Steve Hillyard and his colleagues at the Uni-

Vigilance
An engineer watches the formation of a bar of silicon. High-tech industry demands high levels of constant attention.

Focusing In: Attention and the Brain

1. Erno Rubik
The puzzle cube invented by the Hungarian Erno Rubik (shown here) has become one of the most popular challenges to attention and logical thought.

2. Kendo
Kendo, or the way of the sword, requires harmony of movement and thought. It is inspired by the most famous Japanese book, Treatise of the Five Rings.

3. Tea ceremony
Japanese tea ceremony is based on principles of Zen Buddhism and emphasizes the admiration of beautiful aspects of everyday life. Participants must focus on performing each act according to strict rules of etiquette.

versity of California, San Diego, have found some evidence of this sensitization in minute electrical signals generated by the brain and recorded from the scalp. These event-related potentials occur within a few hundredths of a second after the presentation of the sensory stimulus and reflect the brain's processing of sensory information.

Hillyard's initial experiments required a listener to pay attention to one ear to detect a change in the frequency of tone beeps. A distinctive electrical signal appeared on the scalp when the target beeps arrived at the attended ear but not when the same beep occurred at the unattended ear. This result reflects selective activation of the brain's responsiveness. Later experiments showed that as it becomes more difficult to distinguish the target sound from irrelevant sounds, the characteristic electrical signal occurs later and later, which indicates that there is a slowing of the selection process. Further experiments with event-related potentials have shown that there are different mechanisms for selective attention to locations in space, to color, to orientation, and to form, just as there are different brain regions devoted to responding to these attributes.

Observations of cerebral blood flow by Per Roland at the Karolinska Institute in Stockholm further support the notion that brain activation is associated with paying attention. In his experiments people were told to be prepared to detect an almost imperceptible touch to an index finger. Focusing attention on the finger was accompanied by increased blood flow to the sensory cortex responsive to the finger and to the frontal lobes. The blood flow to the frontal lobes is interesting because a number of studies, particularly with brain-damaged patients, have shown that the frontal lobes prevent attention from being diverted by irrelevant stimuli.

Studies of brain-damaged patients have also revealed other important aspects of attention control. Damage to the parietal lobes at the rear of the brain leads to a failure to pay attention to the side opposite the damaged hemisphere, which receives its sensory information from the opposite side of the body. Patients with damage to the right parietal lobe, for example, have reduced eye movements to the left, draw pictures that ignore the left sides of things, and may even fail to dress the left sides of their bodies. A surprising finding is that

Focusing In: Attention and the Brain

1. Absorbed in combat
In this watercolor from the end of the eighteenth century, two players focus their attention on a game apparently related to the Japanese game of go.

2. Chess
Attention may be diverted by an eruption of violent feelings, such as anger, one of the seven deadly sins illustrated by Collin de Plancy (1864). Unable to control himself, Charles V breaks up a chess game with a monkey who could not be beaten.

flashes of light presented to such patients in their left field of view elicit an event-related potential from the patient's brain even though the patient denies seeing the light. Apparently, the lesion prevents the stimulus from reaching conscious awareness.

Michael Posner of the University of Oregon has intensively studied the process of covert shifts of attention: changing the location in space to which you are paying attention without moving your eyes. His work shows that there are three distinct components to this process. First you must disengage your attention from its current focus. Second, you must move your attention to the new focus, a process that Posner has shown is faster than moving your eyes. Finally, you must engage your attention at the new focus. Posner's work has shown that parietal-lobe lesions disrupt the disengaging process, which makes it difficult for patients with such lesions to let go of something on which they are already focused if the new focus is further toward their neglected side. Lesions in a midbrain cell cluster that receives information from the eyes result in the loss of saccadic eye movements and slow, but do not eliminate, one's ability to make covert shifts of attention. Thus, the midbrain is involved in the control of both movement of the eyes and the movement component of covert shifts of attention.

We would like to know the cellular details of how the brain selects only certain information through the filter of attention. One landmark experiment that begins to shed light on the nature of this filter was conducted on monkeys by Jeffrey Moran and Robert Desimone of the National Institute of Mental Health. They were able to show that cells in the inferior temporal cortex that receive information from the primary visual cortex are affected by attention. Animals were taught to pay attention to a location in space and respond when a particular colored bar appeared there. A bar that would normally cause a burst of electrical activity in a brain cell it appeared in the focus of attention produced little or no response if it fell outside of this focus. This result demonstrates that the brain does indeed filter out irrelevant information and points the way for finding out how such filters may act as sentries controlling the admission of information from the senses to conscious awareness.

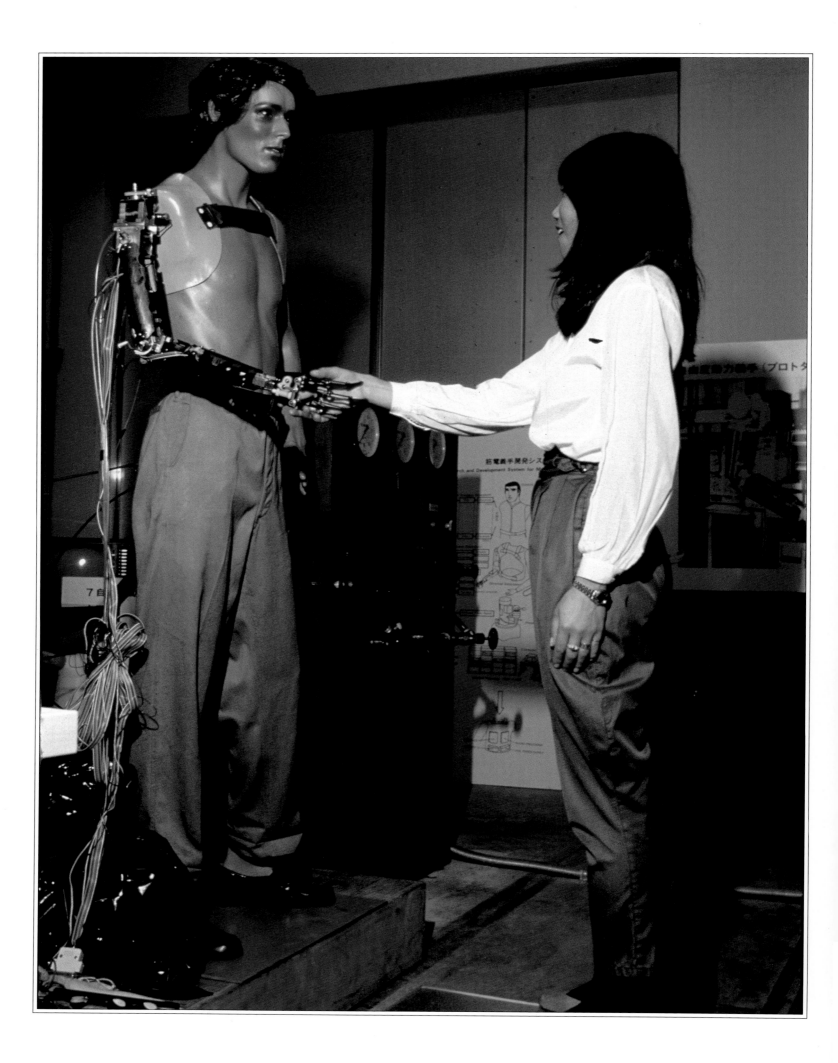

18
Thought as Computation: The Future of Brain Research?

A discussion of computation and thought brings up once again the question of consciousness, which continues to perplex brain researchers and philosophers and more recently has entered the once wholly mechanistic domain of computer science. The question of what consciousness might be is still far from being answered, largely, it seems, because we still do not know precisely how to ask that question. A debate over whether computers can be conscious is thus as empty as the arguments about the number of angels dancing on the head of a pin. At the moment, though, the computational approach to understanding brain function is the best one we have, and it is still far from running dry. Philosophers and neuroscientists alike look to it to buttress cherished old ideas and contribute new insights into the too often overwhelming detail of what the brain is doing.

The idea of massive parallel processing (breaking a problem down into many pieces and working on all of those pieces at the same time) is one of the current great hopes for modeling the brain's operations. The idea seems to hold for functions like vision. Many events are obviously taking place at the same time when we see something, so the subdivisions of the visual system seem to be operating in parallel. But consciousness is clearly a serial operation, capable of being aware of only one thing at a time or at most a few.

The connectionist model

Computer scientists like Jerome Feldman of the University of Rochester propose that we can think about the brain in terms of connections among information-processing units that behave like neurons. The models that he and other brain theorists propose are termed "connectionist models." They study neuroanatomy, neurophysiology, and behavior and suggest that ordinary serial computers are just not sensible models for how the brain works. They point out that brain cells work at the speed of a few thousandths of a second, but that complex behaviors like recognizing printed words happen within a few hundred milliseconds. This means that only about 100 sequential nerve events can happen within that limited time. But if we try to program a standard computer to do such a job, it takes millions of steps. In this view, serial "artificial intelligence" is a blind alley.

Feldman has recently undertaken the formidable challenge of developing a fairly complete abstract connectionist model of how the visual system works. It attempts to account for the ability of primates to locate and categorize objects in space and to use knowledge of the visual environment to guide movement. Moreover, his aim is to develop a model that will be computationally adequate and consistent with what is known about the biology of the visual system and visual behavior. His theory is quite complex and can be sketched here only in very general terms.

The underlying supposition of the model is that four frames of reference are necessary for the brain to convert information from the eyes into information usable to the seer. The first, referred to as the retinotopic frame, relates to what falls on the retina and is presumed to lie in the primary visual cortex. The second is actually a set of frames referred to as stable feature frames. They deal with such features as hue, shape, and size, which are fixed attributes of objects even though they vary as objects move. These presumably lie in the secondary visual areas. The third frame is an integrative frame that brings the observer's cumulative knowledge of the world to bear on the visual scene, along with information from the fourth frame, the information about total visual environment not currently visible to the observer.

The other critical notion in the model is that of the computational units analogous to nerve cells. They have a very large number of incoming and outgoing connections, and the connections among them can vary in their strength in a way that makes learning possible.

From human to robot?
Factories without workers, machines that understand speech and converse, move about, draw, read a musical score: what is the limit?

Thought as Computation: The Future of Brain Research?

1-2. A contest of brains
Solomon Stone (1), the greatest mental calculator in the United States in 1890, and Wim Klein (2), who worked at CERN in the division of theoretical studies until 1977. Klein's prodigious gifts as a calculator were used in parallel with those of a computer. But already the computer excels humans in the domain of calculation. Will it ever be able to excel in creativity?

3. Simulation
Computers make it possible for people to construct objects on the screen before actually manufacturing them. Here a drinking glass is visualized by the Computer Assisted Concept Fabrication system from Computervision. Could a machine design a glass without human intervention?

4. A robotic pianist
Conceived and perfected by Professor Ichiro Kato, this robot was the star of the 1985 Tsukuba exposition. It can read music at a distance of one meter in near real time. It played as it decoded the printed score. Its hands had the same reach as those of humans, and it could accompany a human singer.

5. Adept 1
The robot chooses a chocolate from a box, using a video system controlled by a program that enables the robot to take account of changes that its own actions bring about.

6. The world's largest robot
Another star of Tsukuba, this 25-ton giant, more than five meters tall, was created by Fujitsu.

7. A robotic portraitist
This skilled robot foreshadows the future third-generation robots.

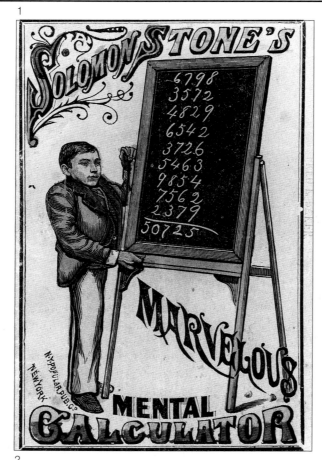

Units in different frames are linked together in complex and changeable ways that permit computations whose results are useful to the observer. For example, information about the size of a person known to an observer compared with the size of the person's image on the retina permits a computation of the distance of the person from the observer. Feldman's model permits far more complex computations, many of which fit with known aspects of neurophysiology and psychology. In addition, it makes certain predictions about what results new behavioral or physiological experiments will yield. Its ability to generate new research questions should make Feldman's model a subject of considerable controversy in the years to come.

Despite the complexity of connectionist models, stripped to its bare bones, the underlying idea of connectionism looks rather skinny and frail. In the brain, each nerve cell is connected to thousands of other nerve cells. Each receives inputs from thousands of others and sends its outputs to thousands, including some that contribute to its inputs. Connectionism is basically just that, a multiplicity of connections and nothing more. But the apparent power of the idea becomes clear when you hear a connectionist model made up of rather ordinary electronic components learn to read. Terry Sejnowski of the Salk Institute and Charles Rosenberg of Johns Hopkins University are the inventors of this device, which can raise goosebumps on the skin of even the most technologically blasé listener. This box full of wires that learns to read is made even more chilling by the childish voice that goes with the show.

But is it any more than just a show? Does connectionism constitute the next frontier in attaching intelligence to matter? Are we at last at a point foreseen not too long after World War II, when Donald O. Hebb said that the secrets of the brain lie in altering the strength of connections between brain cells? It is exciting to think so and perhaps to realize that the secrets the brain has kept from us for so long have been right in front of our eyes for years. If only we knew what questions to ask!

Thought as Computation: The Future of Brain Research?

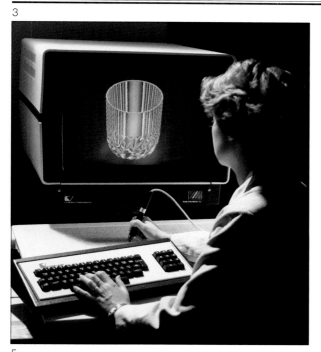

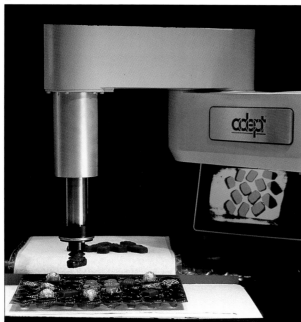

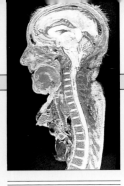

Computers and Brains

by Tomaso Poggio, Artificial Intelligence Laboratory and Center for Biological Information Processing, Massachusetts Institute of Technology

Brains and present digital computers are very different. Brains are made of biological tissues, delicately interwoven cells and fibers. Computers consist of tiny circuits etched in silicon chips and connected to each other by metal wires. Digital computers use only electrical signals; brains use a complex and a not yet fully understood mixture of chemical and electrical activity. While digital computers use binary coding, neurons probably represent information in a more analog form.

Despite these huge differences between hardware and wetware, do computers and brains have anything in common? The answer is yes: both computers and brains process information and have to solve such similar problems in information processing as understanding natural speech or interpreting images. Only recently have researchers realized that this common goal of processing information may lead to fruitful interactions between brain science and computer science.

The computational approach to the brain sciences that has now emerged has its origins in the field of artificial intelligence (AI). The main goal of AI is to develop intelligent artificial systems and, more important, to understand the principles of intelligence itself. Research in AI relies on the principle that information-processing tasks can be studied independently of the hardware that carries them out. This distinction arose naturally: during their short history, computers of different types have been built from a variety of very different components, including relays, gears, tinker toys, lenses, and transistors. In this perspective, neurons can be viewed as simply another type of computing machinery. If intelligent processes can be separated from the hardware on which they run, then intelligence can be studied in its own right. A science of information processing is thus developing that studies fundamental properties of specific information-processing tasks.

An important method for carrying out this computational approach is provided by another fundamental tenet of AI. In AI theories must be tested on a computer to prove that they represent an adequate model from the point of view of information processing. In the computational approach to the brain, the first condition a theory must satisfy is that it must be capable of solving the problem. Computer programs are a critical test for any computational theory. Implementing algorithms on a computer can decide whether the theory is right or wrong as a theory of the computational task; it often reveals features of the problem, or of the solution, that cannot be grasped otherwise.

One of the important contributions of the computational approach thus far has been to demonstrate just how difficult it is to solve problems that our brain routinely solves. This is especially true for vision and control of movement. Precisely these most common abilities of humans and animals seem hardest to understand from the information-processing point of view. Quite appropriately and somewhat ironically, AI systems may replace a lawyer or physician more easily than a gardener or a cook.

As an example of the computational approach, consider the field of vision, one of the most fruitful areas of AI exploration. The study of vision has already established a strong bridge between AI and the experimental neurosciences. Work in computer vision has led recently to a general framework for most of the first steps of vision, or early vision. The goal of early vision is to provide a map of distances and orientations of the surfaces around the viewer from the raw-image data, which consists of large arrays of numbers provided to a computer by a digital camera and represents the light intensity at each pixel (picture element) in the image. Thus, early vision is inverse optics, because it attempts to retrieve the original three-dimensional surfaces from their two-dimensional projections (the images).

We have recently recognized that vision problems are mathematically ill posed. This intrinsic property of vision problems dictates that they should be solved by bringing to bear *a priori* information about generic properties of the solution. Examples of these *a priori* constraints are that objects are typically rigid and that surfaces are mostly continuous with rare boundaries of discontinuities. Constraints of this type must be used by any system, either artificial or biological, that attempts to solve the problem. They are dictated only by the nature of the problem and by its mathematics and physics, not by the properties of the computing machinery.

A specific example within vision is stereopsis. The comparison of the two images taken by the two eyes (or by the cameras of a robot) contains information about the depth of different objects. Stereopsis is the process that extracts this information. Computer vision has recently developed stereo algorithms that work satisfactorily on real images and mimics many of the properties of human stereopsis. Does that mean that computational theories of this type can tell us where and how neurons in the brain are connected to compute depth? The answer is clearly no. A computational theory of stereopsis by itself cannot predict the detailed properties of the cortical neurons that are tuned to specific depths. It cannot tell us how they are connected. It cannot even predict which particular algorithm should be used. This is because an algorithm depends not only, of course, on the problem to be solved but also at least as much on the properties and limitations of the available (and in this case biological) hardware.

In conclusion, the computational approach to neuroscience is *not* an alternative to the more traditional research in the field of neurobiology. Rather, it complements these fields. It is quite clear that AI is already contributing to an understanding of the brain. There is no longer any question of whether powerful computational metaphors such as representations and algorithms are relevant for understanding biological cognition and perception. It is, in fact, quite possible that the computational approach will bring about a new scientific paradigm in the neurosciences, on the same level as molecular neurobiology. On the other hand, computer scientists are also learning from the human brain. Problems in vision and motor control, for instance, are extremely difficult, but biological systems are able to solve them with astonishing reliability and generality. Millions of years of evolution have produced extremely efficient vision systems that can give us useful suggestions about ways to solve certain problems. The cross-fertilization between theoretical questions in information processing, practical applications in robotics, and the natural science of the brain makes the computational approach a most exciting scientific adventure.

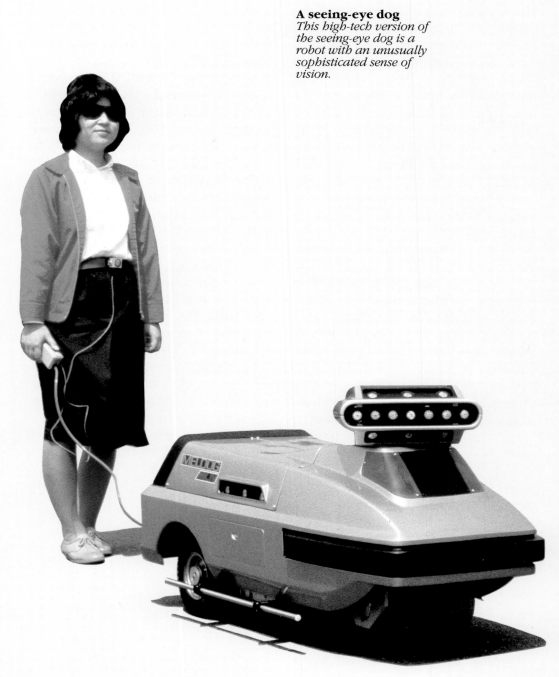

A seeing-eye dog
This high-tech version of the seeing-eye dog is a robot with an unusually sophisticated sense of vision.

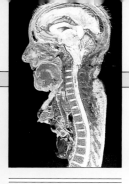

Brains and Computers

by Valentino Braitenberg, director of the Max Planck Institute for Biological Cybernetics

Is the brain a computer? What kind of a computer is it, if it is one? When I am asked these questions by a journalist, I scrutinize her or his facial expression to see if it is one of fear or amused curiosity mixed with incredulity. If it is fear, I can discern various components to that fear: (1) fear that the brain, being something like a computer, is completely determined (as a good machine should be) by the forces that act on it; (2) fear that we are no longer entitled to speak of the mind as something unanalyzable or of the brain as something almost as opaque but are now faced with the overwhelming task of working out a complete blueprint of a totally mechanical brain; (3) the fear of competition: if brains are like machines, there are machines like brains and these may docilely take over such human endeavors as politics, journalism, etc.; (4) the fear of machines rebelling against humans. All of these feelings often add up to a recommendation that we drop brain research altogether, at least the part of it that entails the implicit threat of a perfect analogy between brains and computers.

The brain scientist's answer to such anxiety is, of course, reassurance. One cannot assert, he will say, that the human brain is determined as long as nobody knows enough about it to be able to make even the most trivial predictions about what it will do in a given situation: whether it will organize a sneeze when a certain amount of snuff is fed into the nose or whether it will be able to solve a certain mathematical problem or be willing to come along with me for a walk. This means, of course, that for some time we will continue to talk of a mind with its feelings, wishes and decisions, doubts, and self-awareness, as we always have, not because we fail to see the light of a new cybernetic science that is about to replace these words with formulas of computer engineering, but because the old terms are still the most appropriate ones for the matter at hand, a psyche.

But I do occasionally try to seduce these fearful people a little in the direction of my own tastes. I proceed as follows. I tell them that the central nervous system of a worm is something like an electronic control system and not much more than that. They do not object: nobody cares about worms. Then I tell them that within the cranium of a fly we found a perfect little calculating machine, and we know what it does too: it is the automatic pilot that takes in information from the eyes and calculates the fly's only apparently erratic trajectory from these data. Most people find this interesting and not in the least unsettling, for it is not generally believed that the fly makes his flying turns by high-level moral decisions. Then I go on to say that we have looked at the components of this automatic pilot and have photographed their structure down to the level of the molecules and found them to be fairly complicated but quite understandable in principle, understandable at least enough to fully explain from their compound behavior the workings of the automatic pilot. Now this approach may smack of irreverence toward the Divine Engineer who made living matter, and it may give some people the creeps they have when reading science fiction, but it is acceptable to most people. Then I spring the trap: when we looked at bits of the human brain with the same instruments, light and electron microscopes, microelectrodes, biochemical analyses, etc., we found quite the same components and never anything else, and again we got the feeling that we could fully understand them in principle. Is our brain, then, an automatic pilot, only bigger than that of the fly and perhaps more complicated? My fearful philosophers get the point: yes, it is. But very little is gained by showing the way to analyze down to the molecules, which are the same in flies and men, and very little is lost when we convince ourselves that wonderfully complex affairs owe their existence to very humble building blocks. It is the way things are put together that counts.

To the other kind of people, whose questions about computers and brains stem from true curiosity, I offer a few observations. It is remarkable how similar animal brains and present-day computers are in many ways, even if their operation is radically different, since they are based on different principles. First the similarities. Computers and brains have a fibrous structure: wires or other metallic conductors in one case, processes of nerve cells in the other. Both rely on a precise traffic of signals along predetermined routes: none of the broadcasting "to whom it may concern" that we have in radio communication and in some biochemical systems. Also, the nature of the transmitted signals is strikingly similar in digital computers and brains, short pulses of electric current in both cases, and in both cases the amplitude of the pulse is less critical than the time of its occurrence. This makes for another similarity, namely, the nature of the logical operations (conjunction, disjunction, implication, and the like) that are the function of the operational units in which pulses cooperate to produce more pulses. If in some instances we find this logical description inappropriate for both computers and brains, there are again similarities in the principles of analog computation at work in both cases: frequency modulation, replacement of quantities by their logarithms to facilitate computation, etc.

But in spite of these similarities, I must object when some people stress the point that brains are like computers. They are certainly not like the computers on the market today. They may be very similar to a future generation of computers, and for a good reason: principles of nervous integration discovered by brain scientists are slowly seeping into the world of engineering and have already produced quite a ferment there. Brains have a tendency to use a very large number of their computing elements in parallel, letting them interfere with each other to produce in a very few steps a result that present-day digital computers often reach only after a long chain of many individual operations. Strangely, the slow and slimy components of animal brains, organized in a phalanx of parallel operations, sometimes (e.g., in visual perception) reach their results more speedily than the digital computer, with its vastly quicker metallic and crystalline elements. Once the advantages of parallel computation are incorporated in systems using quick and reliable electronic components, the comparison between brains and computer will become really worthwhile.

Conclusion

What we as human beings do is primarily the result of what our brains do. Understanding ourselves means understanding our brains. The point of the many different stories of research contained in this book is that we are making real and substantial progress toward that goal. But as this book also makes clear, that progress is fragmented and is occurring on many different fronts. Molecular biologists for their part are gaining the ability to look at the communications systems used by neurons at a level that makes possible for the first time a molecular view of the formation of those communications systems from the earliest stages of life and the alterations in those systems that occur during learning and memory formation. At the same time researchers in human language use are beginning to clear away the fog of abstract linguistic theory to understand what the brain is doing when it generates or perceives language.

Not surprisingly, there is often a great deal of misunderstanding between linguists and molecular biologists, between researchers who monitor the electrical activity of individual brain cells and those who interrogate the brain through the medium of electrical signals recorded from the human scalp. Scientists argue over what is "real" science at least as fiercely as artists argue over what is "real" art. Those who like cubism will continue to sneer at the impressionists, and those whose tastes run classical will sneer at Andy Warhol. But the sciences of the mind and the brain cannot afford the luxury of what some may regard as matters of taste. Somehow the molecules that brain cells use to communicate with one another play a role in the equally astonishing process of childhood language learning. We can hope eventually to understand this now rather remote association only if we deeply understand the details of both processes. Molecular biology on its own sheds no light on questions of the mental capacities necessary for language, just as the tools for studying the mental capacities of apes are of no use for identifying the genes that code for neurotransmitters. Yet both provide critical answers to different forms of the question of how the brain works. Each answer is vastly less meaningful without the other; some might even say meaningless.

Is there a potential guiding and integrating influence that can help overcome interdisciplinary squabbles and provide some sense of common goals for psychologists, invertebrate neurophysiologists, molecular biologists, linguists, and other members of the motley crew that man the scientific ship exploring the mind and brain? The rudder may actually be a discipline that many scientists have considered to be dormant, without practical value, or both. The discipline in question is philosophy. It appears from the work of philosophers like Daniel Dennett of Tufts University and Patricia Churchland of the University of California, San Diego, that the field is neither dead nor wanting in practical value to researchers.

Dennett worries about such problems as the ability of living creatures to gain access to relative knowledge stored in their heads in rapid and efficient ways. This is not a matter of idle philosophical speculation, for such access is essential to survival. It is of little value to an organism to sort through information related to food sources in a situation in which mating behavior is called for or to sift through information on potential mates when a predator comes into view.

Dennett sets the problem in the context of artificial intelligence. He challenges researchers to program a robot to use a store of knowledge to survive a dangerous situation: retrieving a needed new battery from a room that also contains a time bomb. Dennett's detailed analysis of the possible difficulties the robot could encounter by being unable to access useful knowledge in a timely manner (for example, that both battery and bomb are on a wagon) illuminates how little we know of common sense and how people manage not to behave like fools most of the time. Nervous systems have evolved to carry out this very puzzling operation without much effort that we can detect. We as humans can access our store of knowledge in very efficient ways, without having to read it out like rules in a handbook. That simplistic notion of how the mind retrieves knowledge stored in the brain is obviously wrong. We need better ideas of what to look for, better questions to ask.

Churchland favors an approach to asking questions about the brain that she terms "neurophilosophy." This is an approach that she as a philosopher has developed through an intensive study of research on mind and brain. She advocates advancing our knowledge of how the operations of the brain bring about the functions of the mind by eliminating what she refers to as "folk psychology" as a source of ideas about what the mind does, effecting an integration of the many disciplines concerned with mind and brain, and bringing to the study of their workings more and better theories that can be tested in the laboratory.

This is not an easy prescription to follow. Folk psychology is deeply ingrained in all of us and hides behind what we often refer to as common sense. Churchland defines it as follows:

Now by folk psychology I mean that rough-hewn set of concepts, generalizations, and rules of thumb we all standardly use in explaining and predicting human behavior. Folk psychology is commonsense psychology – the psychological lore in virtue of which we explain behavior as the outcome of beliefs, desires, perceptions, goals, sensations, and so forth. It is a theory whose generalizations connect mental states to other mental states, to perceptions, and to actions. These homey generalizations are what provide the characterization of the mental states and processes referred to; they are what delimit the "facets" of mental life and define the explananda.[1]

Folk psychology provides us with stories of what the mind does that seem particularly difficult to explain in terms of what the brain does. Some scholars say that this is because such explanation is an impossible task. Churchland says it is because folk psychology is wrong.

The rather vague notion of self-awareness is a good example of folk psychology. Descarte's pronouncement "I think, therefore I am, suggests that he too was misled by folk psychology. Some of the results of split-brain research suggest that individuals do not necessarily have direct access to their mental functions. The experiments by Libet discussed in chapter 1 reinforce this possibility. So too does research on individuals with "blindsight." These persons have blind spots due to damage to or removal of regions of their primary visual cortex, yet they seem able to process visual information of a crude sort without being aware of receiving it. (Blindsight is not, however, a universally accepted phenomenon.) Other neurological deficits, such as jargon aphasia, in which the brain-damaged person speaks nonsense but is not aware of it, cast further doubt on the folk-psychology version of awareness.

What version of how the mind works will replace the fables provided by folk psychology? No one can yet answer that question or even guess the general form an answer might take. But the answer is sure to be consistent with how the brain works. We have at our fingertips an enormous amount of information about how parts of the brain operate in isolation and a few clues as to how some of these parts interact with others. What we lack is any sort of theoretical framework on which we can hang all these bits and pieces to form a sensible, comprehensible pattern of brain function that accounts for our ability to move, think, feel, express ourselves, and do all the other things that make us human.

Building that theoretical framework is by far the most important challenge at this moment in the history of the sciences of mind and brain. The complexity of the human brain is almost frightening to anyone who seriously takes up the challenge of trying to find out how it works. But that did not deter those researchers who refuse to accept the "never" in the skeptic's assertion that we shall never understand how the brain works. Mathematicians, after all, have begun to refute assertions of never in working on hugely complex problems. The power of the computer has been responsible for something of a revolution in mathematics as to what constitutes a mathematical proof. Over the protest of classical mathematicians, their computer-oriented colleagues have been offering reams of computer printout as proof of various theorems. They are too voluminous for anyone to understand as a whole, yet minute study of their parts reveals no flaws.

NOTES

Chapter 1

1. O. Creutzfeldt, "Neurophysical mechanisms and consciousness," in *Brain and Mind*, Ciba Foundation Symposium, 1979, vol. 69.

The Brain and Medical Traditions

Acknowledgment is due to Heidi Thorer, lecturer at the National School of Acupuncture, for her assistance and documentary research.

Sleep and Dreaming

Reprinted with permission of Hoffmann-La Roche.

Alterations in Brain Function

1. L. R. Young, C. M. Oman, D.G.D. Watt, K. E. Money, and B. K. Lichtenberg, "Spatial orientation in weightlessness and readaptation to earth's gravity," *Science* 225 (1984): 205-208; and D. E. Parker, M. F. Reschke, A. P. Arrott, J. L. Homick, and B. K. Lichtenberg, "Otolith tilt-translation reinterpretation following prolonged weightlessness: Implications for preflight training," *Aviat. Space and Environ. Med.* 56 (1985): 601-607.
2. C. M. Oman, B. K. Lichtenberg, K. E. Money, and R. K. McCoy, "Space motion sickness: Symptoms, stimuli, predictability," *Exp. Brain Res.* 64 (1986): 316-334.
3. L. R. Young, C. M. Oman, D.G.D. Watt, K. E. Money, B. K. Lichtenberg, R. V. Kenyon, and A. P. Arrott, "Sensory adaptation to weightlessness and readaptation to one g: An overview," *Exp. Brain Res.* 64 (1986): 291-298.
4. T. Vieville, G. Clement, F. Lestienne, and A. Berthoz, "Adaptive modifications of the optokinetic and vestibulo-ocular reflexes in microgravity," in E. Keller and D. Lee, eds., *Adaptive Processes in Visual and Oculomotor Systems* (London: Pergamon Press, 1986), pp. 111-120.
5. D.G.D. Watt, K. E. Money, and L. M. Tomi, "Effects of prolonged weightlessness on a human otolith-spinal reflex," *Exp. Brain Res.* 64 (1986): 308-315; and M. Reschke, D. Anderson, and J. Homick, "Vestibulospinal reflexes as a function of microgravity," *Science* 225 (1984): 212-214.

Chapter 6

1. George Mandler, "Emotion," in *The Oxford Companion to the Mind*, ed. R. L. Gregory (Oxford: Oxford University Press, 1987), p. 220.
2. José Delgado, "Animal and human emotionality," *Behavior and Brain Sciences* 5 (1982): 425-427.

Chapter 9

1. Patrick Wall, "Pain and no pain," in *Functions of the Brain*, ed. Clive Coen (Oxford: Oxford University Press, 1985), p. 64.

Hearing

The assistance of M. E. Oliver in preparing the figure and typing the manuscript is gratefully acknowledged. Supported in part by NIH grant PO1 NS13126.

Brain and Motor Control

This research was supported by NIH grants NS09343, AM26710, and EY02621.
1. M. Raibert, "A state space model for sensorimotor control and learning," Massachusetts Institute of Technology Artificial Intelligence Laboratory, AIM; and B.K.P. Horn and M. H. Raibert, "Configuration space control," Massachusetts Institute of Technology, Artificial Intelligence Laboratory, memo no. 458, Cambridge, Mass., 1977.
2. T. Flash and J. M. Hollerbach, "Dynamic interactions between limb segments during planar arm movement," abst. 11th annual meeting, Society for Neuroscience, Los Angeles, Calif., 1981.
3. A. P. Georgopoulos, R. Caminiti, and J. F. Kalaska, "Static spatial effects in motor cortex and area 5: Quantitative relations in a two-dimensional space," *Exp. Brain Res.* 54 (1984): 446-454.
4. P. Morasso, "Spatial control of arm movements," Exp. Brain Res. 42 (1981): 223-227.
5. A. G. Feldman, "Change of muscle length due to shift in the equilibrium point of the muscle-load system," *Biofizika* 19 (1974): 534-538; and A. G. Feldman, "Control of muscle length," Biofizika 19 (1974): 749-751.
6. P.M.H. Rack and D. R. Westbury, "Short-range stiffness of active mammalian muscle and its effect on mechanical properties," *J. Physiol. Lon.* 240 (1974): 331-350.
7. A. G. Feldman, "Functional tuning of the system during control of movement or maintenance of a steady posture, III. Mechanographic analysis of the execution by man of the simplest motor tasks," *Biophysics* 2 (1966): 766-775; N. Hogan, "Mechanical impedance control in asservice devices and manipulators," *Proc. Joint Automatic Control Conference*, San Francisco, 1980; N. Hogan, "Impedance control of industrial robots," *Robots Comput. Integrated Mfg.* 1 (1984): 97-113; and E. Bizzi, N. Accornero, W. Chapple, and N. Hogan, "Posture control and trajectory formation during arm movement," J. Neurosci., 4 (1984): 2738-2744.
8. F. A. Mussa-Ivaldi, N. Hogan, and E. Bizzi, "Neural, mechanical, and geometric factors subserving arm posture in humans," *J. Neurosci.* 5 (1986): 2732-2743.
9. E. Bizzi, A. Polit, and P. Morasso, "Mechanisms underlying achievement of final head position," *J. Neurophysiol.* 39 (1976): 435-444; and A. Polit and E. Bizzi, "Processes controlling arm movements in monkeys," *Science* 201 (1978): 1235-1237.
10. E. Bizzi, N. Accornero, W. Chapple, and N. Hogan, "Arm trajectory formation in monkeys," *Exp. Brain Res.* 46 (1982): 139-143.

Is the Brain Involved?

1. M. Stein, R. C. Schiavi, and M. Camerino, "Influence of brain and behavior on the immune system," *Science* 191: 435-440.
2. J. R. Kalden, M. M. Evans, and W. J. Irvine, "The effect of hypophysectomy on the immune response," *Immunology* 18 (1970): 671-679.
3. T. L. Roszman, R. J. Cross, W. H. Brooks, and W. R. Markesberg, "Neuroimmunomodulation: Effects of neural lesions on cellular immunity," in *Neural Modulation of Immunity*, ed. R. Guillemin, M. Cohn, T. Melnechuk (New York: Raven Press, 1985), pp. 95-109.
4. G. F. Solomon and A. A. Amkraut, "Psychoneurœndocrinological effects on the immune response," *Ann. Rev. Microbiol.* 35 (1981): 155-184.
5. K. Bulloch, "Neuroanatomy of lymphoid tissue: A review," in *Neural Modulation of Immunity*, ed. R Guillemin, M. Cohen, and T. Melnechuk (New York: Raven Press, 1985), pp. 111-141.
6. K. Bizière, J. M. Guillaumin, D. Degenne, P. Bardos, M. Renoux, and G. Renoux, "Lateralized neocortical modulation of the T-cell lineage," in *Neural Modulation of Immunity*, ed. R. Guillemin, M. Cohen, and T. Melnechuk (New York: Raven Press, 1985), pp. 81-94.
7. Ibid.
8. G. Renoux and K. Bizière, "Asymmetrical involvement of the cerebral neocortex on the response to an immunopotentiator, sodium diethyldithiocarbamate," *J. Neurosci. Res.* 18 (1987): 230-238.
9. R. Adler, "Behaviorally conditioned modulation of immunity," in *Neural Modulation of Immunity*, ed. R. Guillemin, M. Cohn, and T. Melnechuk (New York: 1985), Raven Press, pp. 55-69.
10. H. O. Besedovsky, A. E. del Rey, and E. Sorkin, "What do the immune system and the brain know about each other?" *Immunology Today* 12 (1983): 342-346.
11. Ibid.
12. Ibid.

Chapter 15

1. J. M. Harlow, "Passage of an iron rod through the head," *Boston Medical and Surgical Journal* 39 (1848): 389-393.

Electrical Signs

1. E. Altenmüller, B. Landwehrmeyer, and R. Jung, "Electrophysiological evidence of right parietal dominance of visual-spatial processing tasks," *J. Neurol.* suppl. 232 (1985): 140; R. Jung, "Electrophysiological cues of the language dominant hemisphere in man: Slow brain potentials during language processing and writing," *Exp. Brain Res.* suppl. 9 (1984): 430-450; and R. Jung, E. Altenmüller, and B. Natsch, "Zur Hemisphärendominanz für Sprache und Rechnen: Elektrophysiologische Korrelate einer Linksdominanz bei Linkshändern," *Neuropsychologia* 22 (1984): 755-775.
2. J. Szentagothai, "The 'module concept' in cerebral architecture," *Brain Res.* 95 (1975): 475-496; and W. G. Walter, "Slow potential waves in the human brain associated with expectancy, attention, and decision," *Arch. Psychiatr. Nervenkr.* 206 (1964): 309-322.
3. D. Ingvar, "Serial aspects of language and speech related to prefrontal cortical activity: A selective review," Human *Neurobiol.* 2 (1983): 177-189.
4. R. Jung, "Hirnelektrische Korrelate von Handlungsintention und Grosshirndominanz," in *Das Verhältnis der Psychiatrie zu ihren Nachbardisziplinen*, ed. H. Heiman (Berlin: Springer), pp. 11-14.

NOTES

Developmental Dyslexia

Supported by NIH grants 14018 and 02711 and by the Beth Israel Hospital's Dyslexia Research Grant.

1. F. R. Vellutino, *Dyslexia: Theory and Research* (Cambridge: MIT Press, 1981)
2. A. M. Galaburda and T. L. Kemper "Cytoarchitectonic abnormalities in developmental dyslexia: A case study," *Annals of Neurology* 6 (1979): 94-100.
3. A. M. Galaburda, G. F. Sherman, G. D. Rosen, F. Aboitiz, and N. Geschwind, "Developmental dyslexia: Four consecutive patients with cortical anomalies," *Annals of Neurology* 18 (1985): 222-233.
4. D. Purves and J. W. Lichtman, *Principles of Neural Development* (Sunderland, Mass.: Sinauer Associates, 1985).
5. R. L. Friede, *Developmental Neuropathology* (New York: Springer-Verlag, 1975).
6. P. S. Goldman-Rakic and P. Rakic, "Experimental modification of gyral patterns," in *Cerebral Dominance: The Biological Foundations,* ed. N. Geschwind and A. M. Galaburda (Cambridge: Harvard University Press, 1984), pp. 179-192.
7. N. Geschwind and W. Levitsky, "Human brain: Left-right asymmetries in temporal speech region," *Science* 161 (1968): 186-187.
8. N. Geschwind and P. O. Behan, "Left-handedness: Association with immune disease, migraine, and developmental learning disorder," *Proceedings of the National Academy of Sciences* (U.S.A.) 79 (1982): 5097-5100; and N. Geschwind, "Immunological associations of cerebral dominance," in *Neuroimmunology,* ed. P. O. Behan and F. Spreafico (New York: Raven Press, 1984), 451-461.
9. G. F. Sherman, A. M. Galaburda, and N. Geschwind, "Cortical anomalies in brains of New Zealand mice: A neuropathologic model of dyslexia?" *Proceedings of the National Academy of Sciences* (U.S.A.) 82 (1985): 8072-8074.

Conclusion

1. Patricia Smith Churchland, *"Neurophilosophy: Toward a Unified Science of the Mind-Brain* (Cambridge: MIT Press, 1986), p. 299.

SUGGESTED READINGS

Adelman, George, ed. *Encyclopedia of Neuroscience.* Boston: Birkhuser, 1987.

Blakemore, Colin. *Mechanics of the Mind.* Cambridge: Cambridge University Press, 1977.

Bloom, Floyd E., and A. Lazerson. *Brain, Mind, and Behavior.* New York: W. H. Freeman, 1988.

Braitenberg, Valentino. *Vehicles.* Cambridge: MIT Press, 1984.

Bullock, T. H., R. Orkand, and A. Grinnell. *Introduction to Nervous Systems.* San Francisco: W. H. Freeman, 1977.

Changeux, Jean-Pierre. *Neuronal Man: The Biology of Mind.* New York: Random House, 1985.

Chomsky, Noam. *Knowledge of Language: Its Nature, Origin, and Use.* New York: Praeger, 1986.

Churchland, Patricia S. *Neurophilosophy.* Cambridge: MIT Press, 1986.

Coleman, Richard M. *Wide Awake at 3:00 A.M.* New York: W. H. Freeman, 1986.

Crick, F.H.C. "Thinking about the brain." *Scientific American,* (March 1979), pp. 219-232.

Dennett, Daniel C. *Brainstorms: Philosophical Essays on Mind and Psychology.* Cambridge: MIT Press, 1984.

Edelman, G. M., and Vernon B. Mountcastle. *The Mindful Brain.* Cambridge: MIT Press, 1978.

Fodor, Jerry A. *The Modularity of Mind.* Cambridge: MIT Press, 1983.

Gardner, Howard. *The Shattered Mind.* New York: Alfred A. Knopf, 1975.

Gardner, Howard. *The Mind's New Science.* New York: Basic Books, 1985.

Gazzaniga, Michael S. *The Social Brain: Discovering the Networks of the Mind.* New York: Basic Books, 1985.

Gazzaniga, Michael S. *The Bisected Brain.* New York: Appleton-Century-Crofts, 1970.

Gleitman, Henry. *Psychology.* New York: Norton, 1981.

Gregory, Richard L., ed. *The Oxford Companion to the Mind.* Oxford: Oxford University Press, 1987.

Griffin, Donald R. *Animal Thinking.* Cambridge: Harvard University Press, 1984.

Hubel, David H., and Torsten N. Wiesel. "Brain mechanisms of vision." In *The Brain.* New York: W. H. Freeman, 1979.

Iverson, Leslie L. "The chemistry of the brain." *Scientific American,* (March 1979), pp. 134-149.

Kandel, Eric R. *Cellular Basis of Behavior: An Introduction to Behavioral Neurobiology.* San Francisco: W. H. Freeman, 1976.

Kandel, Eric R. "Small systems of neurons." *Scientific American,* (March 1979), pp. 66-76.

Klima, Edward S., and Ursula Bellugi. *The Signs of Language.* Cambridge: Harvard University Press, 1979.

Kuffler, S. W., J. G. Nicholls, and A. R. Martin. *From Neuron to Brain: A Cellular Approach to the Nervous System.* 2d ed. Sunderland, Mass.: Sinauer, 1984.

Land, Edwin. "The retinex theory of color vision." *Scientific American* 237 (1977): 108-128.

Lieberman, P. *The Biology and Evolution of Language.* Cambridge: Harvard University Press, 1984.

Llinas, Rudolfo R. "The cortex of the cerebellum." *Scientific American,* (January 1975), pp. 56-71.

Lumsden, Charles J., and Edward O. Wilson. *Genes, Mind, and Culture.* Cambridge: Harvard University Press, 1981.

Marr, David. *Vision: A Computational Investigation into the Human Representation and Processing of Visual Information.* San Francisco: W. H. Freeman, 1982.

McEwen, Bruce S. "Interaction between hormones and nervous tissues." *Scientific American,* (January 1976), pp. 48-58.

Nauta, Walle J. H., and Michael Feirtag. *Fundamental Neuroanatomy.* New York: W. H. Freeman, 1986.

Poizner, Howard, Edward S. Klima, and Ursula Bellugi. *What the Hands Reveal about the Brain.* Cambridge: MIT Press, 1987.

Popper, Karl R. and John C. Eccles. *The Self and Its Brain.* Parts 1 and 2. Berlin: Springer-International, 1977.

Pribram, Karl H. *Languages of the Brain.* Englewood Cliffs, N.J.: Prentice-Hall, 1971.

Pylyshyn, Zenon. *Computation and Cognition: Toward a Foundation for Cognitive Science.* Cambridge: MIT Press, 1984.

Restak, Richard M. *The Brain.* New York: Bantam, 1984.

Sagan, Carl. *The Dragons of Eden: Speculations on the Evolution of Human Intelligence.* New York: Random House, 1977.

Simon, Herbert A. *The Sciences of the Artificial.* Cambridge: MIT Press, 1969.

Snyder, Solomon H. *Biological Aspects of Mental Disorder.* New York: Oxford University Press, 1980.

Squire, Larry R. *Memory and Brain.* New York: Oxford University Press, 1987.

Stillings, Neil A., Mark H. Feinstein, Jay L. Garfield, Edwina L. Rissland, David A. Rosenbaum, Steven E. Weisler, and Lynne Baker-Ward. *Cognitive Science: An Introduction.* Cambridge: MIT Press, 1987.

Wanner, Eric, and Lila Gleitman. *Language Acquisition: The State of the Art.* Cambridge: Cambridge University Press, 1982.

Wolfe, Jeremy M. *The Mind's Eye.* New York:

INDEX

A

Acetylcholine, 98-99, 102-103, 155-156, 167
Acupuncture, 165
AD. See Alzheimer's disease
Ader, Robert, 186
Adey, W. Ross, 173
Adrenaline, 156
AI. See Artificial intelligence
Altman, Joseph, 157
Alzheimer's disease, 90-91 (figs.), 98-99, 100 (figs.), 98-99, 159, 167
Amaral, David, 171
Amphetamine, 156
Anderson, John, 201
Anesthesia, by hypnosis, 183, 186
Anxiety, 94-97
Arendt, Josephine, 66
Artificial intelligence, 223, 226-227, 232
Asanuma, Hiroshi, 138
Aschoff, Jurgen, 63
Attention, 219-221
Auditory system. See hearing
Australopithecus, 26
Autism, 150
Axon, 150, 155

B

Backlund, Erik-Olaf, 159
Basal ganglia in motor control, 140-141, 144-145
Basal ganglia in Parkinson's disease, 156
Bauman, Margaret, 150
Behavior therapy, 34, 36
Behaviorism, 33-38
Bellugi, Ursula, 211
Benzodiazepine. See Tranquilizers
Besedovsky, Hugo, 186
Beta-blockers, 156
Bicameral mind, 13
Biological clocks, 63, 66, 68, 71-73, 75
Björklund, Anders, 159
Blalock, J. Edwin, 186
Blindsight, 233
Blood flow, brain, 16-19
Blood-brain barrier, 86-87
Brain damage, 195-201
Brain damage and attention, 220-221
Brain damage and emotion, 81-82
Brain imaging with computer aided tomography (CAT), 109
Brain imaging with magnetic resonance imaging (MRI), 110
Brain imaging with positron emission tomography (PET), 109
Brain imaging with xenon-133 technique, 16-19
Brain stem, involvement in rapid-eye-movement sleep, 75-76
Broca, Paul, 195
Brodmann, Korbinian, 107
Brown, Roger, 209
Bulloch, Karen, 186

C

Cancer of the brain 198-199 (figs.)
CAT scans. See Brain imaging
Catecholamines, 156
Cerebellum in autism, 150
Cerebellum in movement control, 140-141, 144-145
Cerebellum, learning in, 144, 171-172
Cerebral cortex, 16-19, 22-23, 28
Cerebral cortex in movement control, 135-138, 140-141, 144
Cerebral dominance, See Cerebral hemispheres
Cerebral hemispheres, independence of, 12-13
Cerebral hemispheres, dominance and handedness, 205-207
Cerebral hemispheres, dominance of in Japanese and occidentals, 54-55
Cerebral hemispheres, dominance of in males and females, 45, 60-61
Cerebral hemispheres, separated. See split brains
Chase, Michael, 68
Chemistry of the brain, 155-169
Chomsky, Noam, 210-211
Churchland, Patricia, 232-233
Circadian rhythms, 63, 71-73
Circuits, neuronal, 160-161, 175-178
Cochlea, 131-132
Computers in brain research, 223-230
Conditioning, 172-173
Conditioning, operant, 36-38
Connectionist models of brain function, 223-224
Connections among brain cells, 147, 152-153, 171, 175-178
Consciousness, 11-18, 176-177, 223
Consciousness in split-brain patients, 200-201, 203-204
Consciousness, altered states of, 89, 91
Creutzfeldt, Otto, 11
Crick, Francis, 70
Cricket brain and mating behavior, 39-40
Critical period in development, 176-177
Culture, effects on brain organization, 45-46, 54-55

D

Das, Gopal, 157
Death of brain cells, 147, 150, 152-153, 175
Delay reduction theory, 37-38
Delgado, José, 8
Dementia, 16-19 (*see also* Alzheimer's disease)
Dendrite, 150, 155
Dennett, Daniel, 232
Depression, 190-191
Depression and light, 63
Desimone, Robert, 221
Development of the brain, 147-153, 174-178
Diamond, Marian C., 22
Dopamine and Parkinson's disease, 156-157, 159, 167-169
Dopamine and schizophrenia, 86
Dreams, 68, 70, 74-76
Drugs, 156 (*see also* Tranquilizers)
Drugs, antipsychotic, 89 (fig.)
Drugs and depression, 191
Drugs and dyslexia, 216
Dunnett, Stephen B., 159
Dyslexia, 212-216

E

Ebbeson, Sven O. E., 25
Eccles, John, 12, 155
EEG. See Electroencephalogram
Einstein, Albert, 22-23
Electroencephalogram, 21, 68, 74-75, 216
Electromagnetic fields and brain function, 173
Electroshock therapy, 191
Emotions, 81-83
Emotions and stress, 165
Endorphins, 156, 160, 163
Epilepsy, 197 (fig.)
Epinephrine, 156
Ethology, 33-38
Event-related potentials, 216, 220
Evoked potentials, 21-22
Evolution, 46
Evolution of the brain, 25-27
Exercise and endorphins, 165

F

Fantino, Edmund, 36-38
Feldberg, W., 155
Feldman, Jerome, 223-224
Fenwick, Peter, 89
Festinger, Leon, 25
Filter theory of attention, 219
Fine, Alan, 159
Fixed action patterns, 36
Fliers, E., 45
Flinn, Edward A., 22
Flinn, Jane M., 22
Fodor, Jerry, 201
Folk psychology, 233
Freed, William J., 159
Freeman, Walter J., 122
Freud, Sigmund, 68, 70, 183, 200
Fritsch, Gustav, 135
Fromkin, Victoria, 201
Frontal lobes, 16-19, 91, 195-196

G

Gage, Phineas, 195
Gall, Francis Joseph, 107
Gamma-aminobutyric acid, 94-97, 156
Gardner, Howard, 21
Gazzaniga, Michael, 197, 200
Georgopoulos, Apostolos P., 137-138
Geschwind, Norman, 213
Gleitman, Lila and Henry, 210
Glia, 22-23
Glia, radial, 150
Gödel's incompleteness proof, 11
Grafts, brain. See Transplants, brain

H

Hair cells, 130-132
Hamburger, Victor, 147, 152
Handedness, 205-206
Hearing, 122-123, 130-133
Hebb, Donald O., 224
Hemispheric dominance. See Cerebral hemispheres
Hilgard, Ernest, 183
Hillyard, Steve, 220
Hippocampus and Alzheimer's disease, 98-99
Hippocampus and epilepsy, 197 (fig.)
Hippocampus and Parkinson's disease, 167
Hitzig, Julius, 135
Hoffer, Barry, 159
Homo sapiens, 27
Homunculus, 11
Hormones and behavior, 36
Hormones and brain organization in males and females, 45-46
Hormones of the brain, 162-166
Hubel, David, 118
Humphrey, Donald, 136
Hypnosis, 183, 186
Hypothalamus and brain hormones, 162-163
Hypothalamus and emotions, 82-83
Hypothalamus and immune responses, 192-193
Hypothalamus, sex differences in, 45, 60
Hypothalamus as site of biological clock, 66, 71

I

Immune system, links with the brain, 186, 192-193
Implants, brain. See Transplants, brain
Imprinting, 34, 178
Ingle, David, 118
Intelligence, 21-23

J

Jackson, Hughlings, 135-136
Jet lag, 63, 66
John, E. Roy, 172-173

K

Kimura, Doreen, 45
Kirsch, Arthur D., 22
Klima, Ed, 211
Kosslyn, Stephen, 15
Kwan, H. C., 136

INDEX

L

Land, Edwin, 117-118
Language learning, 34
Language, Japanese, 54-55
Language, 209-216
Lasley, William, 36
Learning and brain development, relation between, 174-178
Learning, connectionist view of, 223-224
Learning disabilities, 212-216
Lehrman, Daniel, 36
Lemurs, 28-29
Letorneau, Paul, 147
Levi-Montalcini, Rita, 147, 152
Libet, Benjamin, 13
Limbic system in autism, 150
Limbic system and emotions, 81-83
Livingston, Margraret, 118
Lobotomy, 91, 195-196
Lorenz, Konrad, 34
Losee-Olson, Susan, 66
LSD, 156
Lumsden, Charles J., 46

M

Madrazo, Ignacio, 159
Mandler, George, 81
Medical traditions, Eastern, 51-52
Meditation, 89
Melatonin and biological clocks, 66
Melzack, Ronald, 124, 164
Memory, 171-178
Mental illness, 85-86, 89, 91
Mesmer, Franz Anton, 183
Modules of the brain, 200-201, 203-204
Moran, Jeffrey, 221
Morphine, 164-165
Morris, Don, 183
Motion, perception of, 119-120
Motor programs, 138-143
Motor systems See Movement, control of
Movement, control of, 135-145, 157
MPTP, 139 (fig.)
MRI. See Brain imaging
Muscles, springlike behavior of, 143
Music, perception of, 54-55

N

Nauta, Walle J. H., 108
Nerve growth factor, 147, 152-153
Neuroanatomy, 107-110
Neuroimmunology, 192-193
Neurophilosophy, 233
Neurotoxins, 102-103
Neurotransmitters, 86, 155-169
Neurotransmitters and anxiety, 94-97
Neurotransmitters and Alzheimer's disease, 98-99, 159, 167
Neurotransmitters and Parkinson's disease, 156, 159, 167-169
Norepinephrine, 156
Nottebohm, Fernando, 45

O

Olfactory system, 120-122, 125-129
Olson, Lars, 159
Orne, Martin, 183

P

Pain, 123-124, 164-166
Parkinson's disease, 101 (fig.), 156, 159, 167-169
Parkinson's disease, drug induced, 139 (fig.)
Pattern generators, neural, 157
Penfield, Wilder, 140
Peptides, 156, 162-163
Perception, sensory, 117-133
Perl, Edward R., 124
Perlow, Mark J., 159
Pert, Candace, 156
PET scans. See Brain imaging
Pheromones, 125-129
Phillips, John, 36
Philosophers' views of brain research, 232-233
Phonotaxis, 39-40
Pigs, smell and reproductive behavior in, 128-129
Piracetam, 216, 217 (fig.)
Pituitary gland, 162-163
Planum temporale, 213
Posner, Michael, 221
Psychiatry, 190
Psychoneuroimmunology, 186

R

Ramachandran, Vilayanur S., 118-119, 201
Rapid-eye-movement sleep, 68, 74-76
Receptors for neurotransmitters, 86, 94-99, 102-103, 155-156
Regeneration, neural, 167-168
Reinforcement, 36-38
REM sleep. See Rapid-eye-movement sleep
Retinex theory, 117-118
Ritvo, Edward, 150
Roland, Per, 220
Rosenberg, Charles, 224
Rosenzweig, Mark R., 22
Rosnethal, Norman E., 63

S

Sack, David A., 63
Schacter, Stanley, 81
Schizophrenia, 85-86
Seiger, Ake, 159
Sejnowski, Terry, 224
Senses. See Perception, sensory
Serotonin, 156
Sex differences in brains, 60-61, 63, 153
Sign language, 211
Skinner, B. F., 33-34, 36
Skull, 26-27
Sleep cycle, 63
Sleep, 68, 74-76
Smell. See Olfactory system
Snyder, Solomon, 156
Sociobiology, 46
Sokoloff, Louis, 89 (fig.)
Somatostatin, 162-163
Space motion sickness, 78-79
Sperry, Roger, 167, 196
Split brains, 196-197, 200-201, 203-204
Squire, Larry R., 171
Stein, Marvin, 186
Stereopsis, 227
Stimulation of the brain, 82-83, 136
Stimulation of the cochlea, 132-133
Stimulation, transcutaneous nerve, 165
Stress, response to, 82-83, 165
Stroke, 202 (figs.)
Swaab, D. F., 45
Synapse, 107 (fig.), 155

T

Temperament, biological basis of, 71-72
Thompson, Richard F., 171
Tinbergen, Niko, 34
Tixier, Jacques, 25
Tomography. See Brain imaging
Toniolo, Guy, 159
Tool use, 26-27
Touch, sense of, 123
Trajectory of movement, 142-143
Tranquilizers, 96-97, 156
Tranquilizers and biological clocks, 66, 68
Transplants, brain, 156-157, 159, 167-169
Treisman, Anne, 219
Trophic factors, 147, 150, 152-153
Turek, Fred, 66

V

Vale, Wylie, 82
Vestibular system, 78-79
Visceral brain. See Limbic system
Vision, 117-120
Vision, binocular, 28-29
Vision, computer, 226-227
Vision, learning and, 174-176

W

Wall, Patrick D., 124
Watson, John B., 33
Weightlessness, effects on brain function, 78-79
Wernicke, Carl, 195
Wild children, 209-210
Wilson, Donald M., 138
Wilson, Edward O., 46
Woolsey, Clinton, 135-136
Wyatt, Richard J., 159

Z

Zelki, S., 118
Zola-Morgan, Stuart, 171

239